T0375073
OCEAN BESTIARY
RJ KING 2022

Ocean Bestiary

Oceans in Depth

A SERIES EDITED BY KATHARINE ANDERSON
AND HELEN M. ROZWADOWSKI

Also by Richard J. King

Lobster
The Devil's Cormorant: A Natural History
Ahab's Rolling Sea: A Natural History of Moby-Dick

Ocean Bestiary

Meeting Marine Life from Abalone to Orca to Zooplankton

Written and illustrated by

RICHARD J. KING

THE UNIVERSITY OF CHICAGO PRESS

CHICAGO AND LONDON

The University of Chicago Press, Chicago 60637
The University of Chicago Press, Ltd., London

Published 2023
Printed in the United States of America

32 31 30 29 28 27 26 25 24 23 1 2 3 4 5

ISBN-13: 978-0-226-81803-0 (cloth)
ISBN-13: 978-0-226-82580-9 (e-book)
DOI: https://doi.org/10.7208/chicago/9780226825809.001.0001

Library of Congress Cataloging-in-Publication Data

Names: King, Richard J., author, illustrator.
Title: Ocean bestiary : meeting marine life from abalone to orca to zooplankton / written and illustrated by Richard J. King.
Other titles: Oceans in depth.
Description: Chicago ; London : The University of Chicago Press, 2023. | Series: Oceans in depth | Includes bibliographical references and index.
Identifiers: LCCN 2022040088 | ISBN 9780226818030 (cloth) | ISBN 9780226825809 (ebook)
Subjects: LCSH: Marine animals.
Classification: LCC QL122 .K54 2023 | DDC 591.77—dc23/eng/20220822
LC record available at https://lccn.loc.gov/2022040088

♾ This paper meets the requirements of ANSI/NISO Z39.48-1992 (Permanence of Paper).

For Alice Day, the brave, the beautiful, the brilliant

Contents

Series Editors' Foreword

Oceans in Depth

What is a bestiary? In classical and medieval times, the bestiary described both familiar and exotic creatures, combining natural history with folklore and moral lessons. A medieval bestiary will tell us that the wolf, for example, is a predator of the earth and sometimes the sky, strong in shoulders and jaw, who cunningly approaches its prey upwind. In the dark, its shining eyes are strangely beautiful, an allegory for the temptations of the devil (https://bestiary.ca/index.html). Richard J. King's bestiary contributes to a lively tradition of writing about our human relationship to the natural world. The ancient bestiary's patterns of knowledge and wonder survive in new ways as tales for our times, with collections of beastlore in nature writing, in literature, music, and popular culture. Amid contemporary environmental damage to oceans, it seems fitting that King should take the bestiary to sea. What lessons might we draw now from a new alphabet of curiosities, recording a world still so mysterious to us?

Because of their mysteries, the oceans have long inspired experimental forms of nature writing. For example, William Beebe, Rachel Carson, and Sylvia Earle, all prominent twentieth-century scientists and authors, wrote about oceans dramatically in an effort to make an inaccessible environment and its foreign forms of life more visible and familiar. King's technique is the exploration of firsthand accounts, telling stories from many different oceans and time periods about the men, women, and children who became absorbed in the strangeness of ocean life. As the voices multiply—Victorian naturalists and modern marine biologists, experienced sailors and novice passengers, whalers and environmentalists—readers gain a rich picture of changing relationships with the marine world. To unsettle our normal categories of nature observer, King even plays with nonhuman voices, imagining an interview with a sea urchin, the suffering of a captive beluga, and the viewpoint of a seagoing parrot, companion to the first Black captain of the US Coast Guard. We become more attentive to the role of human culture, as well as animal agency. *Ocean Bestiary*, then, is above all a history of how we have come to know about the oceans.

Literary scholar and artist, King illustrates how deeply our relationships with the marine world are embedded in both history and imagination. His entries present oceans as sites for historical examination of human encounters with sea creatures. The stories of these meetings leverage the past to offer empathetic perspectives on oceanic environments today and in the future. Some accounts chart devastation, such as the overhunting of sea lion populations or the threats to Pacific Islands from sea-level rise. Others record discovery and adventure, like that of the three generations of marine biologists who gradually learned the secrets of the reclusive paper nautilus. Still others enlist literature, art, or biology to show how our views of ocean creatures have been transformed: the octopus, for instance, shifts from savage monster to reclusive intelligence. King's alphabet, from abalone to zooplankton, provokes curiosity about our continued encounters with the vast and inaccessible marine world. This set of connected tales conveys rich understandings of oceans, leaving readers to ponder the lessons of his *Ocean Bestiary*.

Katharine Anderson
Helen M. Rozwadowski

Introduction

Mr. Thomas L. Albro went to sea as a steward, where he worked mostly in the captain's cabin to serve food to the officers and to clean up, wash laundry, and help sail the ship when others were busy. He seems to have come from an established New England colonial family and perhaps dropped down the social ladder a bit for some reason or another. Or at least he was not ambitious. Maybe, I like to imagine, he was more inclined toward the arts and nature, just happy to serve others if they paid him on time and left him alone so he could draw and gaze over the rail to watch the waves. He went on at least two whaling voyages out of Newport, Rhode Island in the 1830s. The slim historical records that include his name then suggest he returned home, worked as a gardener, married, helped raise children, and at some point moved to an island community in Narragansett Bay where in his old age he was a trustee for the local public school.

The only reason Albro is remembered at all today is because of the artwork that he and his much younger brother created when they sailed together. Their scrimshaw, engraved drawings on sperm whale teeth, are now some of the most prized and valuable possessions in maritime museums around the world. One of Thomas's scrimshaw teeth recently sold at an art auction on Nantucket Island for $102,000.

Another of his pieces of scrimshaw art, known in some small circles as The Albro Bestiary Tooth, is as far as historians can tell entirely unique in what he drew on its surface. It is, in my view, one of the most fascinating pieces of folk art ever left by a sailor anywhere, anytime. The Albro Bestiary Tooth is currently held under glass at Mystic Seaport Museum in Mystic, Connecticut. It looks like a lot of polished, illustrated scrimshaw teeth from the 1800s (as well as their subsequent counterfeits). It is 8⅜" (21 cm) long and was once in the lower jaw of a large male sperm whale that used his teeth for the primary purpose of grabbing and maneuvering slippery squids to slurp down his gullet. If you somehow convinced one of the fastidious curators to unlock the case and give you white gloves to pick up The Albro Bestiary Tooth, you'd find that it is surprisingly hefty, like a long, thick jar.

A careful examination of Albro's engravings all the way around the tooth reveals that the drawings are divided into three separate horizontal scenes.

Two of the scenes are accurate depictions of hunting and processing whales, one of which includes a profile illustration of French Rocks in the Kermadec archipelago in the far southwestern Pacific. This is where the whale itself was harpooned and killed by Albro's shipmates when the animal surfaced after diving for perhaps over a thousand feet, in his search for food. These two whaling scenes include Albro's ship, the *John Coggeshall*, with men high aloft looking for whales and men in boats trying to harpoon the whales who had the fatal misfortune of emerging at this spot for a breather. In these whaling scenes, Albro also drew albatrosses, their wings characteristically enormous and rigid, wheeling over French Rocks.

Thomas L. Albro's third scene, however, is what makes this scrimshaw so extraordinary and unique. Here he inscribed an entire bestiary, a menagerie, a little aquarium of marine life. He drew a large full profile of a sperm whale, and then a line of animals underneath, labeled: a "Pilot Fish" (a small striped fish that likes to hang around boats and sharks), "Dolphin" (meaning a dolphinfish, or mahi-mahi), "Blackfish" (a pilot whale), and "Bill-

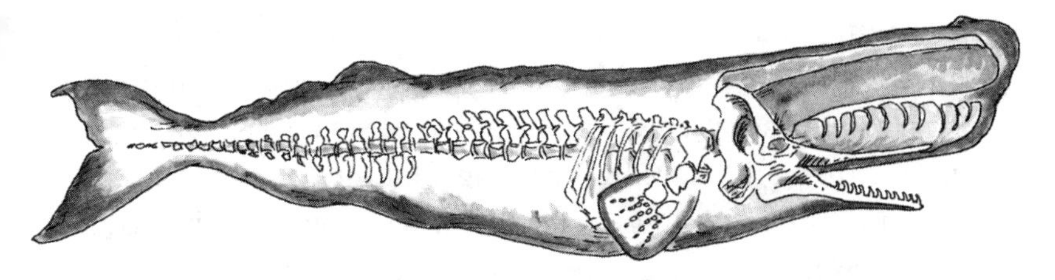

Skeleton of a large male sperm whale (*Physeter macrocephalus*), showing in the head the spermaceti organs and the teeth in the lower jaw.

fish" (a swordfish or marlin). In another line Albro etched animals labeled "Flying Fish," "Albacore," "Porpoise," and "Shark." In between these two lines of animals he engraved a long curvy "Sea Serpant," which was a sea snake. Last of all, unlabeled, he engraved a large sea turtle.

Albro's animal pictures are simple drawings. Yet each has small details that show his careful knowledge, such as the ridges near the tail of the tuna, the long dorsal fin of the dolphinfish, and the left-oriented spout hole of the sperm whale. Not only did his etchings strive for a semblance of accuracy, but when you consider them carefully, you realize that Albro arranged them all to his understanding of food chains, lined up left to right from prey to predator. Albro also did not draw fantastical beasts or animals from distant seas. He did not engrave kraken or mermaids or man-eating lobsters. He did not draw a narwhal. Instead, all of the marine life depicted on his tooth were animals that he surely saw for himself on his Pacific voyages and that he, working with the cook and the other sailors, likely helped capture from the deck of his whaleship then served to the officers and sailors aboard for food. (Except for the sharks; whalers in the 1800s often killed sharks purely out of hatred and to keep them away from their toes and the whale carcasses, often torturing the sharks and returning them maimed to die in the water.)

Sailing in the 1830s, Thomas L. Albro stood on a ship's deck during a time in human history when on average over ten thousand people each year were cruising the ocean on whaleships all around the world, traveling on voyages for three to five years at a time and often spending several months without touching land. There has never been another era on Earth, nor do I think there ever will be again, when so many humans were out at sea quietly, patiently, intently watching the surface of the ocean. Hour after hour, day and night, everyone on the ship was looking, listening, even smelling the air in the hopes of seeing, or hearing, or catching the whiff a whale. Once back ashore, the whalers were keen to show off their firsthand knowledge. They wanted to outshine naturalists on land who thought they were the more reliable authorities about the open ocean, even though those lubberly scientists had worked primarily from reports and pickled specimens they received secondhand. Albro, for his part, drew his bestiary on a trophy, a souvenir, from an animal that he would have watched from a distance. As steward, he stayed on the ship while his brother and others in small boats harpooned and lanced this particular whale until he bled to death in agony. After the men butchered the whale's blubber from the carcass lashed alongside the ship (as they stabbed at the sharks), the sailors then heaved the whale jaw on the deck where they pried out the teeth using ropes and pulleys.

We don't know if Albro had mixed feelings about the morality of killing whales, but if he did, he would not have been alone, even in the 1800s. Either way, he developed a hunter's knowledge from his own experience out at sea. Although he had beautiful handwriting, we don't know if Albro ever spent any time in schools or libraries before his voyages, and I've yet to find any published book from the period from which he might have copied these drawings to engrave on his bestiary tooth.

Written and illustrated two centuries after The Albro Bestiary Tooth, *Ocean Bestiary* aims to share stories of what people have witnessed and experienced out at sea and along the world's coastlines and weave this into the larger history of our human relationship with the ocean by focusing on individual species or groups of animals, one at a time, from *a* to *z*. *Bestiary* is a word passed down from Greek and medieval texts in which accounts and drawings of animals were assembled in a certain way, often alphabetically. I like to imagine that Thomas L. Albro would like this book for the island school he supported. Most of the stories in *Ocean Bestiary* are from firsthand accounts that range in time from those of the earliest Polynesian navigators all the way to observations from today's pilots of deep sea submersibles. Sometimes it has been instructive to dive into the natural history by way of a famous poem, or a novel, or a painting, since these works of art reflect a cultural perception and can have a significant influence on how we think of a given marine animal today. The ocean offers by far the most fertile space on Earth for life, yet it's the most inaccessible region for humans and the place about which we know the least. So sometimes these stories in *Ocean Bestiary* are about how people have come to learn new things about marine animals.

Whalers, scientists, fishers, merchant mariners, engineers, naval sailors, coast guard personnel, pirates, recreational cruisers, sail racers, ocean swimmers, beachgoers, passengers, forced immigrants, enslaved people, tourists, scuba divers, artists, writers, and everyone who has ever gone to sea or spent some time along its edges has made observations about marine life. *Ocean Bestiary* collects some of these impressions and tells each story as an opportunity to learn more about the animals themselves, the history of our relationship with these organisms, and why we seem to rank some species above others. Some of the stories involve retrospectively shocking attitudes and actions against individual animals or entire populations. As hard as they may be to read, in most cases the stories reveal that we are changing our collective behaviors and relationships for the better, though *far* too slowly and at too small a scale.

Not only is the why people went to sea significant to their view of marine life, but the when and the who they were are also essential to contextualizing their observations. Indigenous sailors and the earliest global fishers made the first careful and long-term observations of marine life. Unfortunately, the majority of artistic and historical written responses to the ocean published in English-language records of animals at sea that remain available today were left almost entirely by white men. *Ocean Bestiary* tells stories derived from some of the most famous of these men without delving here into their complicated lives, views, and often horrific results of their voyages, because their environmental accounts are so rare and useful. This bestiary includes accounts left by, for example, John James Audubon, a slaveholder, and by Christopher Columbus, whose voyages directly led to the genocide of Indigenous peoples and cultures. *Ocean Bestiary* also tells stories from white men like Thomas Albro, who published nothing, achieved no fame in their lives, and whose personal views are now unknown. This collection also intentionally explores out-of-print books, rare manuscripts, oral histories, interviews, podcasts, newspapers, archaeological reports, and other sources to tell the stories of people from various racial and ethnic backgrounds, classes, historical periods, and nationalities who inhabited or visited different coastal, island, and open ocean regions. The book tells stories, for example, of the people of Chinese and Filipino descent who fished for shrimp on the Louisiana Coast; the Black captain Harry Dean, who as a young man captured an enormous silver king in the Straits of Florida; the Mashpee

Wampanoag historian Paula Peters, who crafts community art and stories from quahog shells; and a poet of the Pacific Island nation of Kiribati singing the symbolism of the frigatebird. *Ocean Bestiary* features, too, several famous and lesser-known women who described and studied marine life, notably marine biologists Rachel Carson, Eugenie Clark, Elisa Goya Sueyoshi, and Jeannette Villepreux-Power; writer-adventurers, such as Martha Field, Ann Davison, and Lynne Cox; captains, such as Wendy Kitchell; and the women who went to sea as spouses of whaling and merchant ship captains, such as Mary Chipman Lawrence and Mary Brewster. In other words, one mission of *Ocean Bestiary* is to highlight a range of human observers and their relationships with marine life, both historical and contemporary.

Traditionally, bestiaries sought to teach about a menagerie of animal life as they offered lessons on how we humans should think and behave. We'll never know exactly Thomas Albro's meanings for his scrimshaw bestiary, other than to know that his choices and engravings were not frivolous or random. The Albro Bestiary Tooth is an instructive work of folk art, one that reveals in retrospect human cruelty to sperm whales, but it is also an enamel canvas with authenticity, and whimsy, and humor, as evidenced by his sea serpent's flicking tongue and his "You are here"–style manicule, a cartoon pointed finger aimed at the French Rocks to show the location of where this whale was killed.

So in *Ocean Bestiary* I try to follow Albro's lead, to have fun with these stories of people observing marine life at sea, while also occasionally yarning more broadly about marine biology, maritime history, environmental history, and the currents of ocean conservation. I imagine that I'm offering each of these stories to a class of students like the one in his local school, my own humble drawings on the chalkboard.

ARCTIC OCEAN
NORWAY
London
Sicily
Okido
Har
Azores
Sharks'
Bay
Canary
Islands
Suez
Canal
Red Sea
Gulf of
Thailand
Cape
Verde
Islands
Gulf of
Aden
Maldives
St. Peter and
St. Paul
Rocks
Gabon R.
INDIAN
OCEAN
ATLANTIC
OCEAN
MOZAMBIQUE
Cape of
Good Hope
Crozet
Islands

Ikeq-Smith Sound
ALASKA
Bering Island
Kamchatka Peninsula
LABRADOR
NEWFOUNDLAND
Montreal
Grand Banks
New York
Philadelphia
NOVA SCOTIA
Santa Cruz
Monterey Bay
New Orleans
CUBA
Florida Keys
Great Inagua
Cayman Is
St. Croix
Hawaiian Islands
NICARAGUA
PACIFIC OCEAN
Panama Canal
Tarawa
KIRIBATI
Galápagos Islands
Belém
Callao
Tuamotu Islands
Chincha Islands
Kermadec Islands
North Cape
Juan Fernández Islands
Jervis Bay
Cook Strait
St. Clair
Auckland Islands
Cape Horn
SOUTHERN OCEAN
Neko Harbor
Antarctic Peninsula

Abalone

The first story is about a resilient, inspiring group of fishers and their centuries-old relationship with the abalone, which is a type of marine snail. Globally, abalone is a delicacy. It is a valuable harvest for fishing communities around the world. But for the *haenyeo*, the traditional female divers of Korea's Jeju Island, the abalone is also a source of fear, as dangerous as the sharks of their volcanic coastline. Hundreds of these "sea women" over the last three centuries have died in their efforts to capture abalones. There's a saying among these divers: "Haenyeo live with the coffins on their backs."

To understand why the haenyeo see the abalone as dangerous, it helps to know a bit about this animal's biology and where it lives. Among the fifty or so species of abalones found along the world's coasts, the most common two captured by the fishers off Jeju Island are the Pacific abalone (*Haliotis discus hannai*), *jeonbok* in Korean, and the multicolor abalone (*Haliotis diversicolor*), known as *obunjagi*. With a hard oval shell protecting its body, the abalone's organs center around a single, rock-gripping muscle, known as its foot. Abalones are related to other single-shelled gastropods, like periwinkles and limpets, but abalones are far larger and the most powerful. The red abalones (*Haliotis rufescens*) found off the coast of California grow the biggest, with some individuals measuring over twelve inches (30.5 cm) across. The Pacific abalones that the haenyeo harvest can grow as long as six inches (15.24 cm), while the multicolor abalones are much smaller, at about half that size.

With a fringe of tentacles and a rasping band of tiny teeth around their mouths, abalones feed on algae and seaweeds below the tideline, often under ledges or on boulders. They have a row of large respiratory pores curling near the outer edges of their shells, from which they inhale water to their gills, eject waste, and send out their sperm and eggs. In order to hold on to their rocky substrates, which are often subject to strong currents and swells, and to protect themselves against predators, such as octopuses, eels, crabs, and sea otters, abalones can quickly shut tight and seemingly cement their shells to rocks with super-glue force.

Yet people still try to pull them off—or, better, to nab them before the animals notice and hold on for dear life—because the abalone is a such valuable food to so many cultures. The first documented fisheries for abalones occurred 1,500 years ago in China and Japan, but Indigenous peoples around the world have been harvesting the shellfish for far longer. Archaeologists have found abalones in shell middens on Jeju, dating to about 300 BCE, and

Pacific abalone (*Haliotis discus hannai*) alive (*above*); top and inside of empty shell (*below*).

a midden on an island off the coast of California reveals that people have eaten abalones there for over twelve thousand years.

Abalone shell is not only one of the hardest animal-made substances in nature, but the inside of the shell is lustrously silver and pearly iridescent with greens, blues, pinks, and purples. People have valued these shells for decorations, jewelry, and crafts in several coastal communities where historically they rose to a ceremonial and trade value similar to that of wampum made from quahog shell. The Māori people of Aotearoa New Zealand are famous for their use of abalone, known there as *pāua*. For centuries artisans there have used the shells to make fish hooks shimmer, and they embed pieces of abalone shell into wood to represent the eyes of ancestral figures.

On Jeju Island, fishing has been essential because the soil is poor for farming. In early centuries, it was primarily men who dove for abalone and other underwater prizes. The women back then gathered seaweed near the beach. But in the 1600s and early 1700s, men began to fish farther off shore from boats, were sent off to man warships, and moved away to seek better work and to escape paying tributes. Meanwhile, dried abalone remained part of taxes levied on poor coastal communities. People were even flogged if the abalone was not delivered. So as the men fled, the coastal diving shifted to the women, who were forbidden to leave. There is a line of thinking, too, that women are better equipped than men to withstand the cold, with on average a greater percentage of body fat. Perhaps, too, the Jeju women as a collective were and still are emotionally and physically better skilled at holding their breath for minutes at a time, working collaboratively, diving in pairs, often to depths of more than fifty feet (15 m). As the decades continued on Jeju, women went to work at sea during the day, while men stayed ashore for manual labor. Haenyeo even dove while pregnant and sometimes then brought their infants on the boats with them and nursed in between diving sessions.

In 2019 the dangers of diving for abalones and the life of the haenyeo were depicted in a best-selling historical novel in English, *The Island of Sea Women*, written by the Chinese American author Lisa See. She showed that the ability to capture an abalone is a rite of passage on Jeju. Set mostly in the 1930s to the 1950s, Lisa See described the life of the haenyeo before they began to use wetsuits, fins, or weight belts. For centuries haenyeo wore only light cotton swimwear and dove barefoot. Today, haenyeo, as they did historically, still swim out with a net attached to a *taewak*, a buoy float, and dive with goggles. They swim with a knife for emergencies and strap around one wrist a *bitchang*, a small flat metal crowbar. Sometimes, depending on what they're

Haenyeo diving with a *bitchang*, wearing the clothing used before wetsuits and fins.

gathering, haenyeo dive with a small spear, a hooked tool, or a seaweed sickle.

In *The Island of Sea Women*, the daughter, named Young-sook, remembers her first tragic dive for abalones. She was fishing with her mother, the chief haenyeo: "We approached slowly so as not to disturb the waters. As fast as a snake striking its prey, I thrust my bitchang under the lifted edge of the abalone and flipped it off its home before it had a chance to clamp down. I grabbed it as it started to fall to the seabed. Seeing I was successful and having more air than I did, Mother thrust her bitchang under another abalone just as I started to kick for the surface."

Young-sook surfaces and triumphantly holds up her first abalone. The older haenyeo shout their congratulations. Young-sook, following the custom, rubs the abalone along her cheek to show her appreciation and gratefulness.

Young-sook then realizes that her mother has not yet come up. She dives back down to find that her mother's bitchang is seized by the abalone. Her mother can't get the tool free, and she had dropped her knife trying to cut the strap around her wrist. Young-sook, only about fifteen years old, tries to help. The two fumble with her knife, cutting the mother's wrist. The sea clouds with blood. Her mother kicks and pulls frantically against the abalone's grip. Then she stops. Through her goggles, she looks Young-sook in the eyes to say goodbye before she drowns.

Young-sook returns to the surface without her mother. She lives the rest of her life with the knowledge that she was not only unable to free her, but perhaps her triumphant kick had startled the other abalone to seize her mother's bitchang. Young-sook goes on to be the haenyeo chief, but her life is a hard one. In *The Island of Sea Women*, Lisa See shows Western readers the dangers that the haenyeo have endured in order to provide abalones and other seafood for consumers ashore.

Today, several abalone populations around the world are threatened, largely due to overharvesting, coastal pollution, and climate change. This is especially true off the west coast of North America, where six species were once abundant only decades ago, and now, despite fisheries regulations, at least two are on the brink of extinction, exacerbated by loss of healthy kelp habitat, disease, and competition with urchins. If you're eating abalone at a sushi restaurant, there is a good chance it was raised in an aquaculture facility of some kind.

On Jeju, the haenyeo are and have been sustainable subsistence harvesters for centuries. They have been careful as to where, when, and at what size they gathered abalones and other marine life in order to maintain the stocks for their daughters and their daughters' daughters. Their coastal territories are self-regulated. They shun scuba gear so as not to be tempted to overharvest. They and local managers help release young farm-raised snails and abalones to rebuild the populations, and they clear seaweed in places to reserve the rock substrate for the shellfish. Yet still abalones seem to have all but disappeared from Jeju's coastal waters, perhaps for reasons connected to climate change and run-off pollution.

And the haenyeo themselves are endangered. Listed by UNESCO as an Intangible Cultural Heritage of Humanity, over 90 percent of the women are over sixty years old. In 1965 there were about 23,000 haenyeo fishers on Jeju. By 2019 there were only 1,579 registered divers. Younger generations have sought other jobs where the work is not so hard.

For example, when asked in 2017 how her life as a haenyeo has changed in the last fifty years, Lee Mae-chun said, "Abalone was also easy to spot and to collect, but these days, no more. I still remember the excitement of catching a lot of abalone during one dive. Abalone was very valuable and you could earn good money for it."

Architeuthis dux

The Reverend Moses Harvey was in his house one day in the autumn of 1873 when Tom Piccot knocked on the door. Tom, the twelve-year old son of a local fisher, asked if the reverend would be interested in buying part of a squid tentacle. It was nineteen feet (5.8 m) long, like a thick coil of giant kelp. The tentacle was pinkish gray with a spade-like end clustered with dozens of suckers, each of which had within a hard ring of teeth.

Reverend Harvey stood at the door looking at the tentacle. He was fifty-three years old and had more hair in his beard than on his head. He was known as a lucid speaker, a prolific writer on religion and science, and the person along the coast most interested in any sort of odd animal or plant. That's why the boy has been sent to walk across the peninsula here in Newfoundland and bring it to Harvey's door in St. John's. The reverend purchased the tentacle immediately.

"I was now the possessor of one of the rarest curiosities in the whole animal kingdom," Reverend Harvey wrote years later, "the veritable arm of the hitherto mythical devil-fish, about whose existence naturalists had been disputing for centuries[.] I knew that I held in my hand the key of the great mystery, and that a new chapter would now be added to Natural History."

According to Harvey's final telling of the story, young Tom explained to him that he, his father, and another fisher had been out on Conception Bay when they saw a big mass in the water, poked it with a gaff, and then, to their horror, found themselves enveloped in the arm and tentacle of an enormous squid. The animal began pulling the hull under, sinking their boat. They saw its large, gnashing, parrot-like beak at the base of its arms. At seemingly the last moment, amid the sea spray and the boat leaning steeply, Tom had the good sense to grab the hatchet and hack off the squid's grasping, coiling, sliming arm and tentacle. The animal ejected buckets of musky ink into the sea and receded beneath the surface.

Terrified and shaken, the three fishers hurried back ashore and kept the two chopped off extremities. They were so unaware of the significance of these squid parts to science that they chucked the animal's arm to some dogs on the beach, and the only reason Tom kept the tentacle, roughly half its full length, was because he thought it might be useful as spare rope.

At this time no one in the Western scientific community, on either side of the Atlantic, had anything but a couple stories of variable veracity about giant squids, or giant octopuses, or "devil fish," stories that had been entwined with myths of sea serpents and kraken in various fantastical bestiaries and a few scattered biological illustrations. Beginning in 1870, however, a few more credible accounts of these animals began returning home with fishers from Newfoundland and Massachusetts, men who had been working in coastal waters and on the Grand Banks. Some oceanographic anomaly had compelled a previously unknown number of giant squids (*Architeuthis dux*) to occasionally breach the surface of the western North Atlantic. This happened for several years. Reverend Harvey had been following the phenomena. He pickled his new treasure, Tom's giant squid tentacle, in brine and began writing to experts about it. Harvey had the tentacle photographed and illustrated, and then he gave it to the local museum.

About three weeks later, before Harvey had finished his full writings about the tentacle, he received word that even closer, just to the north in Logy Bay, four fishers had encountered yet another giant squid and managed to capture the entire animal in their herring net. In the struggle to kill the enormous thing with their knives, they ended up stabbing the animal's eyes and severing most of its head and mantle, but they were still able to bring the nearly full carcass ashore, however decapitated. Harvey paid them for the squid on the condition that they cart it to his house. They did so gladly for a royal sum at the time of ten dollars.

Moses Harvey was beside himself.

"My happiness was complete," he wrote.

After soaking the enormous specimen in brine, he laid it out and measured it. This squid's two tentacles measured twenty-four feet (7.3 m) long, while its eight arms each stretched about six feet (1.8 m). In full, this squid measured thirty-two feet (9.8 m) from the tip of the mantle to the end of the tentacles. Reverend Harvey wondered, since this one appeared to be much smaller than Tom Piccot's attacker, if this giant squid had been a sad female who had lost her wits, mourning her lost mate over in Conception Bay. (Biologists have since confirmed that it is actually the females that are larger, and no one still has much of a clue about the mating habits or emotional loyalties of giant squids. The male's prehensile penis, which can extend over three feet long, was only identified recently.)

The tentacles of giant squids can get stretched out, but if Reverend Harvey's Logy Bay squid was indeed about thirty-two feet in life, this would be about average for an adult female. Current biologists believe, on the basis of collected parts and growth rates, that there just might be fifty-foot (15.2 m), or maybe even sixty-foot (18.3 m) giant squids somewhere out there, which was

Reverend Harvey's estimate of the length of Tom Piccot's squid, based on the size of its tentacle.

Many townspeople came over to see the giant squid specimen at Reverend Harvey's house. So many gawkers ambled over that he had to limit his visiting hours. I like to think he and his wife Sarah served tea and biscuits. Meanwhile Harvey and a colleague brought the whole carcass over to a photographer's studio and draped the arms and tentacles over a sponge bath rack with the beak sticking up like an enormous split walnut on top of a giant cupcake.

"The photograph, like George Washington, cannot tell a lie," Harvey wrote. "Had I published the story without the attesting photographs, . . . my story would have been placed on the same level as those about the mythical sea-serpent."

Reverend Harvey sent the nearly complete specimen of the giant squid, presumably by ship and train or horse-cart, sloshing in preservative, and down to Yale University, where it was eagerly received by a young professor with a walrus mustache named Addison Verrill, who had been studying squids and tracking all the recent sightings. With an artist colleague, Dr. Verrill compiled the first scientific paper and the first set of accurate illustrations of a giant squid that had ever been published. Verrill named the specimen *Architeuthis harveyi* in the reverend's honor. (Fortunately, Reverend Harvey died before this scientific moniker was demoted, because the giant squid had by a previous scientist already been named *Architeuthis dux*, which means roughly the "leading first squid.")

Today biologists believe that *Architeuthis dux* spends most of its life in complete darkness between roughly 600 and 3,300 feet (~200–1,000 m) deep. To see in the dark, giant squids have evolved eyes as large as soccer balls to find their prey of fish and

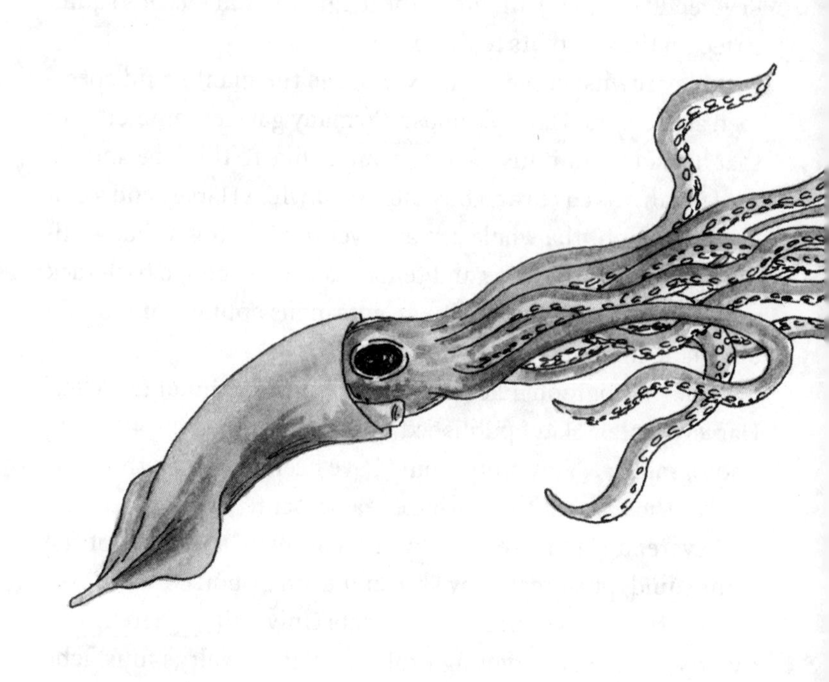

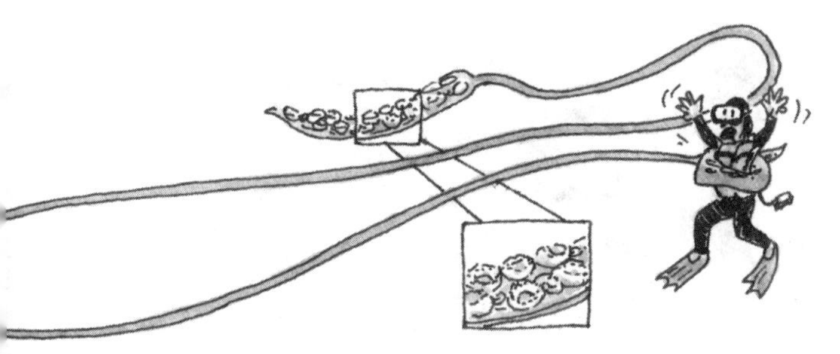

Giant squid (*Architeuthis dux*) with its eight arms and two tentacles.
Inset shows magnification of tentacle suckers.

other squids and to better elude their primary predators, which are sperm whales, other deep-diving toothed whales, and large fish, such as swordfish and tuna. Giant squids are not known to spend much time at the surface at all—unless they're dead—at which point they are usually eaten by sharks and seabirds. This is why that decade of multiple giant squid encounters in the 1870s was so outlandish.

Since the time of Reverend Moses Harvey's first photograph of a full specimen of *Architeuthis dux*, other photographs of dead, beached giant squids have been published on occasion. It was not until 2004, however, that a Japanese biologist named Tsunemi Kubodera was able to capture an image of a *living* giant squid, which he photographed on a baited trip line at nearly 3,000 feet (900 m) down.

The Reverend Moses Harvey declared: "The mists which ignorance and superstition had collected around these animal forms are now dissipated, and they are known in their real proportions and shapes. The mystery has been swept away."

In truth, beyond only a few glimpses of what they look like alive, today we still know barely the first thing about *Architeuthis dux*.

Beluga

This story is not a pleasant one, nor does it have a happy ending, but it teaches about a specific kind of whale, about marine mammal biology in general, and about the way many human communities think of whales has changed dramatically over the last two centuries.

In May of 1877 hunters captured a beluga whale in a net off Labrador, Canada, where these animals at the time were killed commercially for food and oil. This individual, though, was not to be butchered. The Royal Aquarium in England had funded the capture of this beluga, a nine-foot (2.7 m) long female, and her transport across the Atlantic, with the idea of putting her on public display at their site in Westminster. The small whale was laid on seaweed in a coffin-like box. For two weeks she was carried aboard a ship to Montreal. From there she was transported by train to New York, where she was then kept for the summer in an aquarium reservoir beside Coney Island. In September the beluga was again placed in a box on a bed of seaweed and lashed on the deck of a steamship, the *Oder*, which traveled across the Atlantic. During the passage the man who directed her original capture regularly doused her with seawater.

Just as Aristotle and the fishers of ancient Greece had known two millennia before, scientists understood in the late 1800s that marine mammals can survive for quite a long time out of the water as long as their skin is kept moist and their blowhole kept clear to breathe. The men transporting the whale on the *Oder* believed that if they kept the beluga whale in a water-filled tank she would've been injured sloshing around on a ship.

Meanwhile, as the *Oder* steamed across the Atlantic, the staff at the Royal Aquarium built an enormous custom iron pool, forty feet long by twenty feet wide by six feet deep (12 × 6 × 1.8 m). They recessed the iron pool into the ground, filled it with fresh water, and built aisles of seating for visitors to look down upon the tank.

Belugas, which are medium-sized toothed whales, are found in Arctic and sub-Arctic waters. They have often been known as "white whales" or "white porpoises" because of their white skin as adults. (The name *beluga* comes from the Russian, *beloye*, which means white.) Male belugas grow larger than females, over seventeen feet (5.2 m) long. Belugas have no dorsal fin, which might aid their swimming in shallows underneath the ice and help to reduce heat loss. They are exceptionally dexterous in tight shallow spaces, maneuvering in narrow ice leads, capturing quick-moving prey, and escaping predators such as polar bears and killer whales. Belugas can also dive over a half-mile (.8 km) be-

Beluga whale (*Delphinapterus leucas*).

neath the surface as they search for their food of squids, fish, and crustaceans.

When the *Oder* arrived in England, the men placed the box with the beluga on a smaller boat, steamed her to shore, and then placed her on a train and to the aquarium. When the whale arrived, they eased her into the water. She had not eaten a thing during the entire journey from New York.

At first the beluga sank. Unnaturally. The people worried she would drown. In his essay "The White Whale," Henry Lee, a naturalist on the scene, wrote, "But presently up came the beautiful creature; the blow-hole appeared just above the water, the breath was exhaled in a gentle puff, a fresh supply of air being at the same moment taken in, and then the whale descended in a graceful curve, to repeat these movements again and again."

Having survived her long journey overland and then across the Atlantic, the beluga seemed to be doing well. She feasted on the live eels they released into the tank. But three days later the beluga whale died. The eels began to bite into her fins. Yet Londoners still paid to see the corpse.

Perhaps she died from stress or exhaustion. Maybe her internal organs and muscles were damaged, since whales are not built to withstand the pressure of their heavy bodies on a flat surface for such a long time. She probably did not die because of the fresh water, however, because belugas often make their way up rivers, far from the sea and salt water. The staff that examined the animal afterward, which included Henry Lee, found pneumonia, phlegm in her lungs, which they believed dated from about the time her ocean passage began, perhaps because of stress or her inability to regulate her own temperature. Today, professionals that have to transport whales keep the animals partially submerged and suspended in special stretchers in a tank with carefully monitored water temperature.

In telling the story of the beluga, Henry Lee synthesized as much as Western science knew at the time about these animals. (Lee, as it happened in the small world of nineteenth-century marine biology, was also a correspondent with the Reverend Moses Harvey; Lee published his own work dismissing myths about the giant squid.) In the process of their autopsy, the staff found the size of the beluga's brain to be larger than that of a human, prompting them to wonder about the intelligence of these whales.

At another point in "The White Whale," Henry Lee wrote about beluga vocalizations: "The blowing of the Beluga is said to be not unmusical at sea. When it takes place under water it often makes a peculiar sound, which might be mistaken for the whistling of a bird; hence one of the names given to it by sailors—the 'sea canary.'"

The Inuit had long known that the beluga, which they call

qilalugaq qaqortaq, make some of the most varied vocalizations of any marine mammal. From their blowholes, belugas make a range of over fifty different sounds, which include moos, whistles, tweets, buzzes, and a wet, brambling sound that is quite like a human fart. They have individual "signature" calls with which they can identify each other. Echolocating through their oil-filled foreheads, their melons, belugas send out a further variety of clicks, squeaks, and rattles. If you get the chance to visit a beluga at an aquarium, you can hear many of these noises when they break the surface or even underwater through the glass of the tank, although some of the sounds they make are beyond the human range of hearing.

Henry Lee marveled at the white whale's beauty. Yet, at least in his writing on the topic, he did not in the late 1800s consider the capture, transport, or captivity cruel to the animals. Animal-rights groups such as the Royal Society for the Prevention of Cruelty to Animals had been active in England since the 1820s, but the focus was more on horses and dogs, rather than marine life. The death of this beluga did inspire some outcry, including one newspaper that questioned if "to exhibit the whale in an aquarium at all does not amount to something like cruelty." Beginning in 1861, belugas were some of the first whales that aquariums regularly tried to display in captivity in Britain and the United States. Several belugas had been brought down from Labrador to Boston and to New York City. A few individuals survived and thrilled thousands of visitors. More often, though, the whales died in route or died soon after arrival. In 1865, two beluga whales lived in a tank for a short time and then died horribly, boiled to death in a fire at P. T. Barnum's Museum in New York City. New York City aquaria continued to capture and subsequently watch die still more belugas in their efforts to show them to the urban public.

Soon after the loss of the beluga whale at the Royal Aquarium in England in 1877, four more animals were shipped across the Atlantic the following year. One beluga died along the way, suffocating when it got flipped over in its box during rough weather. The remaining three were placed at different public facilities, including one in that tank at the Royal Aquarium at Westminster. After ten days, it too died, but not before over thirty-six thousand people came to lay their eyes on the little white whale. Officials tried to cover up the death by inserting overnight another of the belugas that had recently arrived, but then this one died, too.

Debates about keeping beluga whales in aquaria are active and continue to evolve in the twenty-first century. Even as exhibits are now far larger and healthier for the animals, and the quality of care has vastly increased, no aquarium can mimic a beluga's life out at sea. On the other hand, aquaria in North America and Britain no longer collect marine mammals unless they would die at sea because of poor health. And whales living in aquariums do provide learning opportunities to people who might in turn help wild populations. In 1878 Henry Lee wrote of the beluga's death: "The public, too, were deprived of a great sight, from an educational point of view. Thousands of persons who had opportunities of seeing the porpoises in the Brighton Aquarium learned then for the first time to appreciate the fact that the cetacea are not fishes." For Henry Lee, the educational opportunity to teach anatomy was reason enough.

Most advocates for keeping marine life in aquaria today believe its benefit cannot be solely biological information, but must also be for the purposes of teaching conservation, developing research intended to benefit the animal's larger population and the rehabilitation of individual animals who have been injured in the wild. For example, biologists continue to learn from belugas in aquaria about the behavioral meanings behind their vocalizations, the impact of anthropogenic sound, about beluga pathologies and their longevity: researchers analyzing beluga teeth have found that beluga whales can live to be eighty years old. Much of this research could never have been done in the wild. Scientists have also been developing techniques to study how belugas in the Arctic are responding to global warming. As more ice melts each year, with more year-round open water, beluga whales seem to be faring better than other species, although they are vulnerable to increased human activities in these newly opened waters, such as shipping traffic, fishing, and coastal pollution.

Thus, beluga whales, now safe in most parts of the Arctic and sub-Arctic from capture for aquaria and non-Native hunting, are today an indicator species for how nonhuman marine life will respond to climate change: a sea canary in the coal mine.

Chinstrap Penguin

Once a person was swimming in the waters of Antarctica, on purpose, when she was suddenly surrounded by chinstrap penguins. To explain how this came to happen, it's prudent to first discuss penguins, who spend about three-quarters of their lives in the sea.

Penguins evolved from flying seabirds. Beginning over fifty-five million years ago, their penguin ancestors, which back then were far taller and heavier than our human swimmer, slowly shed their ability to fly in the air, adapting to swimming in cold, deep water. Biologists delineate today about eighteen species of penguins, all of which live in the Southern Hemisphere. Some penguin species swim and breed in more temperate waters, including the endangered Galápagos penguin (*Spheniscus mendiculus*), who lives up near the equator. But even these lower latitude penguins dive along coasts where the water is still numbingly cold, primarily on the west and south sides of continents, like off South Africa and Chile or off islands like those of Aotearoa New Zealand—all places where coastal upwelling insures bracing deep water and a regular supply of fish, squids, and krill. And penguins live in Antarctica, too, of course.

The chinstrap penguin (*Pygoscelis antarcticus*), the type that swam with the person that day, is one of the more common global penguins in terms of numbers and range, and one of a handful of penguin species that raise their chicks in the waters of Antarctica. Perhaps as many as eight million chinstrap penguins migrate each year to the coast of the Antarctic Peninsula and islands such as those of South Georgia, South Sandwich, and the South Shetlands. Chinstrap penguins, also known as bearded penguins, have black backs, black flippers, and pink feet, and they have a unique black cap plumage, like a French beret, and a thin black line under their beaks, which does looks like a chinstrap or a thin beard. Chinstrap penguins are mid-sized among the global penguins, at most 2.5 feet tall (.8 m), weighing up to nearly twelve pounds (5.3 kg).

In other words, seeing chinstrap penguin species on the shores of Antarctica was not a surprise. Chinstrap penguins breed in tight colonies, often with other penguin species, like Adélie and gentoo penguins, although the chinstraps tend to prefer the higher, more precarious nesting spots on the rocky hills, using their beaks and the black claws on their feet to climb up.

The day of the swim was December 15, 2002. Though it was the austral summer, the water was 32°F (0°C), nearing the freezing point of sea water in the region, which is about 28°F (–1.9°C). The extreme athlete was Lynne Cox, attempting to be the first known person to swim in Antarctica—without a wetsuit or flippers or an air tank—just a swim cap, goggles, and a regular one-piece bathing suit. She slipped into the water from a steel gangway on a ship anchored in Neko Harbor on the western, windward side of the Antarctic Peninsula.

Lynne Cox is a mammal, of course, and mammals regulate body temperature primarily through blood flow and a variety of strategies to warm and cool this blood. Without feathers or much fur, our human response to immersion in cold water is that the outer capillaries immediately shrink, forcing warm blood back to the regions of the body where it's critical—the heart, the brain, the internal organs—and then we try to survive the cold by physical activity, frantic or intentional, like shivering and swimming, in order to warm up our inner core as our extremities grow numb and the muscles stiff. Eventually our blood vessels reopen to return some blood to the rest of the body to avoid frostbite. Then we're doing okay in the cold water for a short while. But if our activity can't warm the blood enough, then the vessels can't help but bring the colder blood back into the core, leading to hypothermia and then death, often quite quickly.

Unlike mammals, penguins are evolved for constant immersion in frigid seawater: have a heavy layer of fat to stay warm. They primarily use their feathers as insulation, even, in the case of the chinstrap penguin, at depths of up to two hundred and thirty feet (70 m). Penguin feathers grow densely set on their skin with no bare spots, and the plumage extends over their legs. Notably unlike humans, penguins have a heat exchanging network of blood vessels in their wings and in their feet to warm ingoing colder blood with the outgoing warmer blood. To take care of their insulating plumage, penguins spend hours each day preening, insuring their feathers remain water repellent. Meanwhile,

Chinstrap penguins (*Pygoscelis antarcticus*).

the black plumage on their back and head absorbs any heat from the sun, in or out of the water, and, as with pelagic fish, the black back adds some camouflage in the sea from predators looking down on the swimming penguin, while the white bellies provide camouflage from predators and prey looking up toward the brighter surface.

Penguin wings—more like flippers, really—are supported by flat, partially fused bones that cannot fold and are supported by strong muscles that help them paddle through the water. They use their stiff tails and their feet as rudders and auxiliary propulsion for their streamlined bodies. When swimming at the surface, chinstrap and other penguin species will even leap out of the water, "porpoising," which allows them to breathe and to maintain their speed while confusing and escaping predators. Penguin eyes are adapted to vision underwater, and they have keen hearing to tune in to echolocation or the sounds of

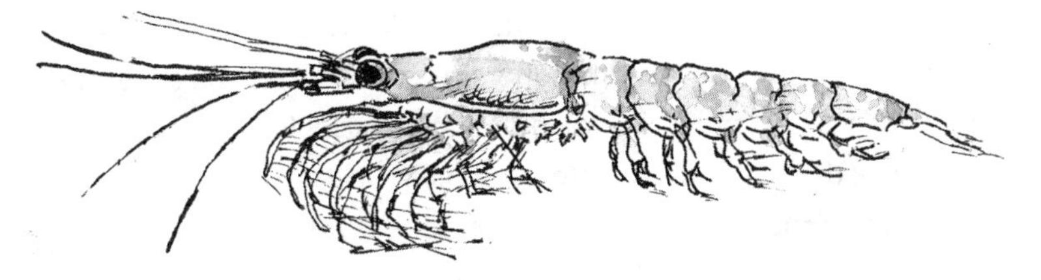

Antarctic krill (*Euphausia superba*), up to 2.5" (6.3 cm).

their predators, marine mammals and sharks. Feeding almost exclusively on krill, chinstrap penguins may swim over thirty miles (48 km) away from their nests to find food for their chicks. When not breeding during the winter, they disperse and swim away from the ice and into the tempestuous Southern Ocean. One recent study of colonies on the Antarctic Peninsula tracked chinstrap penguins traveling as far as five hundred miles away (800 km) from their breeding colonies in the off-season; some of them voyaged as far away as three thousand miles (4,800 km)—that's about the distance across the Atlantic.

Lynne Cox had a few physiological advantages that made her especially good at swimming in cold water. As mentioned regarding the haenyeo, women in general have more evenly distributed body fat, and Cox has strong thighs, big hips, and large breasts. For this particular swim, she had put on more weight, and she grew her hair long to add insulation to the top of her head. Medical researchers have found that Cox's capillaries, unlike those of nearly all other people alive, barely reopen at all during cold immersion—so cooled blood has less chance to recirculate into her head and torso. As Cox has explained, "My body basically says lose the hands, lose the feet, keep the core warm and keep the brain and lungs and heart going."

More significant than any physical advantages, however, Cox has a superhuman drive and mental strength for ocean swimming. She had trained as a distance cold water swimmer since she was a child. At only fifteen years old, she set the speed record, female or male, for the swim across the English Channel. Then she did it again a couple years later. Still a teenager, she was the first woman to swim across the Cook Strait in Aotearoa New Zealand. By the time of her Antarctic dip, among several other record passages, Cox had swum across the Straits of Magellan in water that was 42°F (5.5°C). In 1987 she stroked from one island to another across the international waters of the Bering Strait in water that was 40°F (4.4°C), a swim that was genuinely a major event in the thawing of the Cold War.

After Lynne Cox entered the sea that day in Antarctica, bracing against the burning cold and shock, swimming at first with her head out of the water, she found herself paddling far more rapidly than she wanted. In her trial swim the day before, the cold had damaged, numbed some of the nerves under her skin. She tried to keep her inner core warm. She tried to relax. As she swam, though, she had to maneuver around brashes of ice—occasionally hitting ice chunks with her head. She pawed through water that was nearly slush.

As Cox swam, three doctors, an Antarctic diver in a wetsuit, camera people, and a few of her friends all drove beside her in two inflatable boats, known as Zodiacs, as she swam. The polar entourage were all bundled in parkas, monitoring her every movement and facial expression in case she became disoriented or went into cardiac arrest. Before the swim, the lead doctor had asked to do a trial run of carrying her in a stretcher back up the gangway of the ship. He also insisted on using a marker to draw on her hands the location of her veins in case he had to administer an IV needle in an emergency.

Yet after twenty-five minutes and 1.2 miles (1.9 km), Cox began to head to the beach to the finish. And that's when the penguins came.

"One hundred yards from shore, I saw chinstrap penguins sliding headfirst, like tiny black toboggans, down a steep snowbank," Cox wrote in her memoir *Swimming to Antarctica* (2004). "When they reached the base of the hill," she continued, "they used their bristly tails like brakes, sticking them into the snow to stop their momentum. They waddled across the beach at full tilt, holding their wings out at their sides for balance. Reaching the water, they dove in headfirst, then porpoised across it, clearing it by one or two feet with each surface dive. They tucked their wings back by their sides so they would be more aerodynamic." Approaching the beach among the penguins, Lynne Cox had not only survived, but she had accomplished this first ever Antarctic mile swim, pushing her body beyond anything people thought possible.

She wrote about the chinstrap penguins: "When they neared the Zodiacs, they dove and flapped their wings under the water as if they were flying through air. It was amazing to think this was the only place they would fly. They zoomed under me in bursts of speed, and their bubbles exploded like white fireworks. More penguins joined in. One cannonballed off a ledge, another slipped on some ice and belly flopped, and three penguins swam within inches of my hands. I reached out to touch one, but he swerved and flapped his wings, so he moved just beyond my fingertips. I had no idea why they were swimming with me, but I knew it was a good sign; it meant there were no killer whales or leopard seals in the area."

It took Lynne Cox a full three months to get the feeling back in her flippers.

Dolphinfish

The Norwegian anthropologist-explorer Thor Heyerdahl and five fellow adventurers set out from Callao, Peru, aboard a raft made of logs. They named their raft *Kon-Tiki*. The men were all buddies from the Second World War, and they wanted to go to sea to get away from it all. Heyerdahl had a dubious theory that the human settlement of the islands of the Pacific moved westward from white, bearded people in South America, but it seems that the crew, besides Thor, didn't care too much about the theory; they just loved the idea of going on a huge raft with a great big sail for a very long time.

Heyerdahl and his buddies began their sail on *Kon-Tiki* in April of 1947, a passage that would stretch nearly five thousand miles across the open eastern Pacific and last over three months. Along the way they had a rare view of the ocean, one which was closer to the perspective of the true Polynesian navigators who had in fact settled the Pacific Islands, beginning from Southeast Asia and expanding, fanning out eastward.

Among the marine life that Heyerdahl and his crew on *Kon-Tiki* observed during their long, slow passage across the Pacific were schools of dolphinfish. In his best-selling book *Kon-Tiki: Across the Pacific by Raft* (1950), Heyerdahl wrote about these fish: "There was not a day on which we had not six or seven dolphins following us in circles round and under the raft." Sometimes they recorded as many as thirty or forty dolphinfish swimming underneath *Kon-Tiki* or leaping in their wake.

Heyerdahl was not talking about the marine mammal here, but about the cold-blooded ocean fish called the dolphinfish, also known as mahi-mahi or dorado. There are two different species. The larger and more global one is the common dolphinfish (*Coryphaena hippurus*). The other is the pompano dolphinfish (*Coryphaena equiselis*). Both live in temperate and tropical waters all over the Earth, but the common dolphinfish is more widespread and, well, more common. Dolphinfish of both species are famous for eating flying fish, even leaping out of the sea to nab them midair. Perhaps to help them hold on to these fish in flight, dolphinfish have a lot of small teeth, including a patch of teeth on their tongues.

Male common dolphinfish are usually larger and heavier than the females. The males can weigh over eighty pounds (36 kg) and grow to over six feet (1.8 m) long. The scientific genus name, *Coryphaena*, is a combination of the Greek words for "bright" or "radiant" and for "helmet," since the male dolphin has a particularly square forehead. The female's forehead is more sloped. The Hawaiian name *mahi-mahi* means strong-strong.

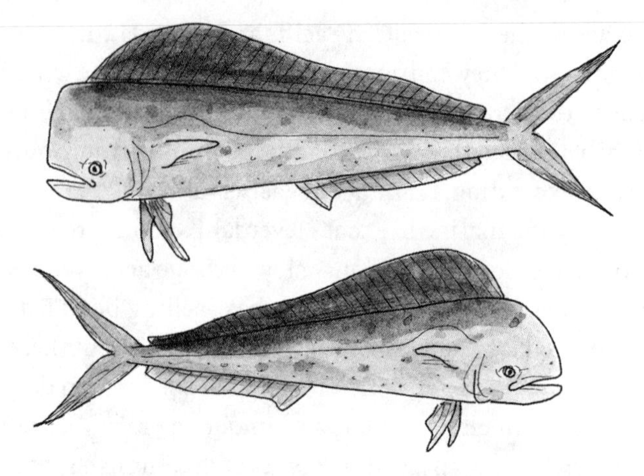

Dolphinfish, or mahi-mahi (*Coryphaena hippurus*): male (*above*) and female (*below*).

Dolphinfish followed Heyerdahl's *Kon-Tiki* because they like to swim under the shade of floating reeds, branches, or any other kind of debris. Knowing this, the Ancient Romans rowed out into the Mediterranean and cast lines under rafts they had constructed especially to catch these fish. They have remained popular food fish for sailors of all kinds on deep ocean voyages.

If you like to eat fish, dolphinfish are considered delicious. They've always been popular with mariners, who have found them to be of the easier species to catch out at sea, usually with a small harpoon or by dragging a line off the stern with a baited, shiny lure and hook. In 1789, for example, these were the first fish that Captain Bligh and his starving crew were able to catch from their small boat in the South Pacific after being cast adrift by the mutineers.

Dolphinfish are perhaps best known for their colors: an iridescent bronze body with greens and purple shades and blue spots. They have a long royal blue dorsal fin and a carmine yellow tail. Poets, sailors, and naturalists have been fascinated by these colors, so the dolphinfish has several cameos in famous sea stories.

In 2001 Canadian novelist Yann Martel in *Life of Pi* wrote how a dolphinfish "shimmered neon-like." Author-naturalists John Steinbeck and Ed Ricketts in their *Sea of Cortez* (1941) wrote of these "startlingly beautiful fish of pure gold, pulsing and fading and changing colors." In 1952, Ann Davison, the first woman to sail alone across an ocean, wrote of dolphinfish, "gorgeous creatures," as her most reliable companions; a single dolphinfish swam over to her little wooden boat each morning and became so familiar that she would touch its nose with a canvas bucket. A century earlier, Richard Henry Dana Jr., in *Two Years before the Mast* (1840) wrote that the fish is "the most elegantly formed, and also the quickest fish, in salt water; and the rays of the sun striking upon it, in its rapid and changing motions, reflected from the water, make it look like a stray beam from a rainbow." The famed painter of birds, John James Audubon, drew a huge dolphinfish in his journal in 1826 when sailing in the Gulf of Mexico. Audubon wrote that these fish gleamed through the waters like meteors, "a blaze of all the hues of the rainbow intermingled." Decades before him in 1788, the English sailor-poet William Falconer described the fish in verse:

> But now, beneath the lofty vessel's stern,
> A shoal of sportive dolphins they discern
> Beaming from burnish'd scales refulgent rays
> Till all the glowing ocean seems to blaze.

As most of these authors and artists and mariners observed, when the dolphinfish dies, the multicolored skin fades to a dull silver, and often dramatically fast. In 1818 the English poet Lord Byron wrote,

> parting day
> Dies like the dolphin, whom each pang imbues
> With a new colour as it gasps away,
> The last still loveliest, till—'tis gone—and all is grey.

Aboard *Kon-Tiki*, as Heyerdahl and his crew sailed across the eastern Pacific, they watched a school of dolphinfish attack a big sea turtle, hunting collectively. They watched dolphinfish leaping into the air after flying fish and schooling and scattering to try to escape sharks and tuna.

"The dolphin had a magnificent color," Heyerdahl wrote. "In the water it shone blue and green like a bluebottle with a glitter of golden-yellow fins."

Although a dolphinfish once bit one of the crewmember's toes when he brought it on deck, Heyerdahl's crew often swam among these fish. They spent hours leaning over the side of *Kon-Tiki* and watching the dolphinfish change colors "like a chameleon." The men regularly hooked the fish for dinner.

After 101 days on the Pacific Ocean in "neighborly intimacy with the sea," the six adventurers crash-landed their raft *Kon-Tiki* onto the coral reef of Raroia in the Tuamotu group of French Polynesia. The waves smashed the raft to splinters, but the men survived. Although the expedition proved little about the history of Pacific migration, they had a lovely time at sea and left us some useful observations of the dolphinfish.

Electric Ray

Eugenie Clark was a pioneer of marine biology. The daughter of a Japanese American mother and a white American father who died before she turned two, Clark grew up in New York City in the 1920s and '30s. From the time she could walk, her mother, Yumico Mitomi, took Eugenie to the beach on Long Island to swim and to eat sushi in downtown Manhattan. Eugenie also went to the public aquarium every Saturday. Soon she convinced her mother to buy her a tank for Christmas, including a few fish and a book on home aquaria.

"Throughout high school, fish were on my mind," Clark wrote later. "No matter what topic we were told to write about for our

English class compositions, I could usually slant the subject to bring in fish."

When Eugenie Clark grew older, she earned her PhD from New York University. In the 1940s, personal scuba technology was still being developed, so she conducted her field studies in a diving helmet connected by a hose to a boat. Or she snorkeled with a wide oval mask, what was then called "underwater goggling." Clark began her research in southern California, then continued in Palau and other islands in the southwestern Pacific. In Guam and Indonesia, Clark learned how to spear fish in open water, as well as to throw traditional fishing nets. She collected and observed fish around coral reefs. She focused on identification and analysis of poisonous fish, such as pufferfish. Soon Clark was featured regularly in magazines such as *Natural History* and *National Geographic*. Her first book, *Lady with a Spear*, was published in 1951 and was the first popular memoir by a female marine biologist. *Lady with a Spear* was an international bestseller, coming out near the same time as *Kon-Tiki* and part of a major postwar movement that brought public attention to marine biology.

One of Clark's most harrowing adventures occurred one day when she was spearfishing in the Red Sea as part of a year-long study of reef fish, working out of a marine lab in Egypt. Clark was snorkeling at the surface in the turquoise, clear waters of Sharks' Bay when she saw at the bottom a large, flat, motionless fish. Using the weight of her heavy spear like a dive belt to help her sink, Clark swam down toward what she hoped was one of the poisonous sting rays she wanted to capture and bring back to the lab for study.

"Just as I was about to thrust the spear in the fish's back, I recognized the species. It was a ray all right—an electric ray!

My spear would have been a fine conductor if the fish and I had connected."

Biologists have identified at least fifty-two species of electric rays around the world, classified in four different families with lovely common and scientific names: the numbfishes (*Narcinidae*), the sleeper rays (*Narkidae*), the coffin fishes (*Hypnidae*), and the torpedo rays (*Torpedinidae*). In *Lady with a Spear*, Clark did not tell her reader exactly which species she nearly stabbed that day, but one of the more common electric rays in the Red Sea is the

Dorsal and ventral sides of a leopard torpedo ray (*Torpedo panthera*). Electrical muscles are under the skin on the dorsal side.

leopard torpedo ray (*Torpedo panthera*), first named by the Western scientific community in the 1830s.

Electric rays have been observed in the Middle East and the Mediterranean for thousands of years, described in accounts by Aristotle and Pliny. Electric rays appear on ancient pottery and in mosaics. In early Greece, patients with headaches were prescribed live electric rays to touch. Cooked electric rays were served to relieve other maladies, such as arthritis. Claudian, a Roman poet from the fourth century CE, wrote: "But nature has armed [the electric ray's] flanks with a numbing poison and mingled with its marrow chill to freeze all living creatures, hiding as it were its own winter in its heart; . . . all [fish] that have touched it lie benumbed."

Beginning in the 1700s, naturalists began using electric rays, along with electric eels and electric catfish, to study the foundations of electricity. Military engineers began to design underwater explosive cartridges, which they named torpedoes. In more recent decades, the analysis of electrified muscles in fish has helped doctors understand the tiny electric pulses that travel through our nerves.

Electric rays range in size. Some species can grow six feet (1.8 m) long. Electric rays in general tend to be more sluggish and more drably colored than other rays, spending their energy on concealment, rather than speed. As with other rays, their gill slits and mouth are underneath, on their ventral side, but they have little gill-like spiracles on top, behind their eyes, to help them breathe when lying flat on the ocean floor and covering their full set of gills.

In the wild, electric rays use their charged muscles to deter predators, such as sharks, and stun their own prey, such as fish, octopuses, and lobsters. Electric rays zap their prey from a short distance away, sending the charge through the water. Different species of electric rays have different amounts of voltage, which they pulse in short bursts from specialized kidney-shaped muscles on each side of their backs. Some of the larger types send out a pop with enough power to light a couple of lamps. This can shock human fishers into dropping their lines or nets. If Eugenie Clark had not checked herself at the last second and had actually harpooned the electric ray with her metal spear, the stun might have knocked her backward in the water.

After that near miss in the Red Sea and her stunning bestseller, Eugenie Clark became more famous for her study of sharks and a range of other fishes' behavior. Clark swam up the ranks to be a master scuba diver, expedition leader, chief scientist, author, advisor, and a professor at the University of Maryland. Dr. Clark founded and directed the Mote Marine Laboratory in Sarasota, Florida, leading it from a single room beach shack to a large, internationally renowned research facility and aquarium.

Eugenie Clark revisited the Red Sea many times after that encounter with the electric ray, advocating for further research and conservation in the region. For her ninety-second birthday in 2015, she returned to the Red Sea for one final dive to her old research sites. She did not carry a spear this time.

Flying Fish

Few animals at sea are as iconic or bring as much joy to sailors as flying fish. The "wings" on which flying fish soar are pectoral fins adapted to spread out as gliders. The long spines of these pectoral fins are connected with a thin, often translucent skin, not unlike a Chinese fan. Charles Darwin considered the flight of these fish in his paradigm-shifting *On the Origin of Species* (1859), imagining flying fish as a transitional group of animals evolving toward full flight. For most mariners, the magical vision of glistening flying fish gliding over the surface when scattered by a boat's bow or fluttering up as a school to avoid predators underneath is a delightful signal of good fortune and fair weather. For Poon Lim, flying fish not only nodded to fair winds: they saved his life.

During the Second World War, Lim, who grew up in Hainan, China, was working as a steward aboard the British supply ship *Benlomond*, bound across the South Atlantic toward Suriname. On November 23, 1942, about 750 miles east of Brazil, a German submarine fired two torpedoes into his ship. Poon Lim was in his cabin.

By the time he got up on deck—amid the alarms, smoke, and screams of burning and trapped shipmates—his assigned lifeboat had already been launched. Lim hurried into his life jacket. He and a couple other men leaped over the rail into the wreckage. The *Benlomond* sank within two minutes. Lim had not swam since he was a child, but he managed to stay afloat by finding a piece of wood, and then a hatch cover. Terrified, he floated past dead shipmates, wreckage, and smoke. His wet clothes dragged him down. His hair and skin were covered in black oil. Eventually he found an empty ship's raft. By the time the sun had set, Poon Lim was floating all alone in the middle of the South Atlantic.

Four decades later, the story of Lim's experience was told by the Chinese American author Ruthanne Lum McCunn. She crafted the narrative, *Sole Survivor*, from newspaper accounts, naval research, and several interviews with Lim and his family. In *Sole Survivor* McCunn wrote that flying fish were the marine life that provided early physical and spiritual sustenance to carry Lim through and give him hope.

The wartime emergency rafts of the Royal Navy had been carefully planned, and though Lim's English was not perfect, he was able to figure out the directions and access a small store of water and food stored in tins, which included hard tack, pemmican (dried, preserved meat and fat), and some chocolate, all of which he was able to stretch out for a while. Early on, while his flashlight still worked, Lim found that flying fish were attracted to the light, flopping onto his little deck. During his first days as a castaway, though, he did not eat the flying fish because he still had food stores—and the idea of raw fish did not appeal to him.

His wood raft was not much more than an eight-foot (2.4 m) square floating dock, constructed with two decks sandwiched around metal drums filled with food, water containers, and supplies. The raft could flip and float either way up. Lim figured out how to slot the four poles into each corner and to rig the canvas tarp, protecting himself from the sun. As the days turned into weeks, he managed to angle the canvas to create a pocket to catch rainwater.

His food, however, even as he ate sparingly, began to run out. The containers had no fishing gear. His raft was often investigated by sharks. Small fish huddled underneath, especially as algae, barnacles, and other fouling organisms began to settle and grow on his raft beneath the waterline. Lim tried to catch the smaller fish with his hands, but was unsuccessful. When the battery to the flashlight died, flying fish only landed on his deck on a bright moon, attracted by the reflection off the canvas. Poon Lim continued to watch the shimmering, silver flying fish with interest and now famished longing.

"Suddenly they leaped," wrote McCunn. "Fins beating, they soared, then landed, then soared again, like the flat stones he used to skip across the stream at home. As they passed [to] starboard, he realized they were being chased, leaping to get out of reach of snapping jaws. Sharks?"

Lim observed long golden-blue fish, likely dolphinfish or tuna, chasing after the flying fish. This was something mariners have observed for millennia, including Aristotle and surely the Polynesian voyagers. In 1726, for example, when Ben Franklin sailed on a trans-Atlantic trip at twenty years old, he not only described tropicbirds and pilot fish, but he wrote carefully about

A dolphinfish hunting a flying fish.

dissecting dolphinfish on the ship's deck and finding flying fish in their bellies. A century later John James Audubon, sailing in the Gulf of Mexico, found twenty-two flying fish, each about 6–7 inches (~15–18 cm) long, packed inside a dolphinfish stomach "like so many salted herrings packed in a box." Only a few years later in the Pacific, the whaler Thomas L. Albro engraved his "Albacore" tuna pointed toward his "Flying Fish."

At first Poon Lim looked at these flying fish with excitement, simply as a natural phenomenon, focusing his energies on trying to catch the fish he could reach for food. He fashioned a little hook out of the spring of a dead flashlight and used barnacles as bait. With this, he managed to catch at least some tiny fish. But these provided little sustenance. Realizing he needed a larger hook, with a key he painstakingly dug out a nail from the raft, hour after hour, finally resorting to using his teeth, his molars, to painfully pry out the nail. Now he put the tiny fish he caught on the larger hook. Soon he was able to catch fish that were three feet long. He dried the strips of meat or ate the fish raw. On desperate days, he chewed on the fish's vertebrae.

Meanwhile, in the rare moments he could ponder the beauty

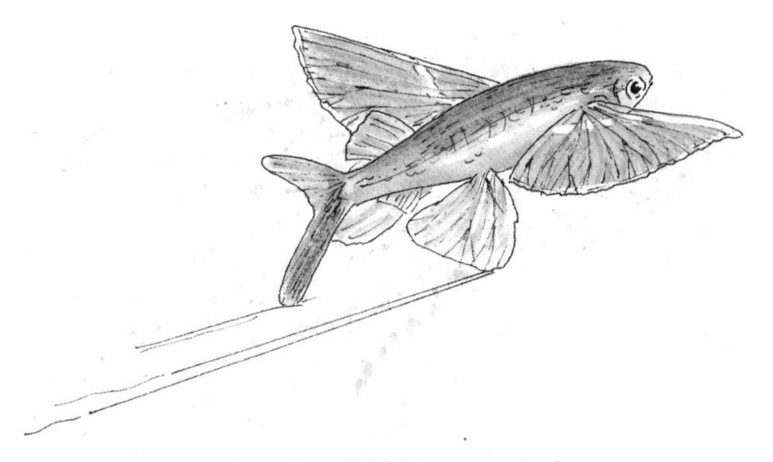

Bandwing flying fish (*Cheilopogon exsiliens*),
a four-winged type, gliding above the surface.

of his surroundings, he watched flying fish skittering through rainbow-lit sea spray. He dreamed of all the flying fish in the sea making for his betrothed a bridge on which to walk from their home village to come visit him out at sea.

With nothing but time, McCunn wrote, Poon Lim watched the flying fish carefully and began to differentiate the species. He observed two and four-winged types and the "beautiful colors." He noted the different bands on their wings, the asymmetrical lobes of their tails and even the "flaplike whiskers" at the tip of some of their lower jaws.

Marine biologists today have identified as many as sixty-seven species of flying fish (family Exocoetidae) swimming and gliding throughout all temperate and tropical waters of the global ocean. They indeed divide the flying fish today between the two- and four-winged types. Flying fish adults can grow from six to twenty inches (15–51 cm) long. They need a substrate on which to lay their eggs, so they usually deposit them in the fronds of sargassum or other floating ocean algae.

McCunn further described Lim's observations of flying fish: "In preparing to leave the water, they would drive very rapidly upward, and if their initial spurts were not sufficient to get underway, they would vibrate the long, lower lobes of their tails sideways, like outboard motors, until they were airborne. Holding their wings steady, they flew close to the water, going 150 feet (46 m) and more in a matter of seconds. Watching them, Lim imagined himself skating across the surface as easily as they did until he reached home."

Most sailors have stories of walking a ship's deck at night and being smacked directly in the face with a flying fish. (I've been slapped by a flying fish on multiple occasions.) Individual flying fish have notched record glides of more than a quarter-mile (.4 km), and, given the right draft of wind, reached heights of over forty feet (12 m) above the surface. Even mariners on ships like the *Benlomond*, whose decks are over three stories above the sea, have found flying fish in their scuppers in the morning.

As the months wore on, Poon Lim ate every fish he could capture, as well as any that flew across his raft. On the 133rd day of his ordeal, a local fishing vessel run by a husband, wife, and their daughter, picked him up off the mouth of the Amazon River. They sailed with him for three days to the city of Belém, Brazil, where he arrived in perfect health. From here Poon Lim was escorted, interviewed, fêted, and brought to London to meet the queen and brief the Royal Navy on how he survived.

Lim returned to work on merchant ships at sea as a steward, and over the years he advocated for the equal rights of Chinese mariners. He retired in Brooklyn, New York, and died in 1991 at the age of seventy-two, a few years after seeing the publication of *Sole Survivor*.

I like to imagine that Poon Lim approved especially of Ruthanne Lum McCunn's description of one dark moment on the raft, late in his ordeal, when he mistook flying fish for a flock of birds: "He slumped back down into the bedding [a fold of canvas], angry at himself for making such a stupid mistake. For the flight of the flying fish was mere imitation, and they did not belong in the birds' sky any more than he belonged in the fish's watery world."

Frigatebird

The ocean territory of the nation of Kiribati extends beyond the coasts of thirty-three islands and rolls across a vast expanse of the equatorial Pacific Ocean that is, west to east, as wide as the United States.

Roughly 112,000 people live in Kiribati today, mostly on the island of Tarawa. Since reclaiming their independence from British and American rule in 1979, the I-Kiribati have adopted a bright red flag design with a blue ocean, a yellow island, and a yellow seabird. This seabird, top and center on the flag, is a frigatebird, the country's national symbol and, in life, is mostly black. The frigatebird, with its thin, sharp wings, is famous in the tropics for extraordinary powers of flight.

A year before independence, an I-Kiribati poet wrote "The Song of the Frigate Te Itei." It is the story of a mother frigatebird who flies off to find food for her young, only to return and find her island entirely gone. Her home has been flooded over. The song ends as follows:

> I have been alone for so long
> Rise up—you, the centre of the world
> Rise up from the depths of the sea
> So, you may be seen from afar
> Rise up! Rise up!

Beyond being a sadly beautiful story on its own and especially moving when performed, this song is significant for thinking about our changing perceptions of the ocean and seabirds.

"The Song of the Frigate Te Itei" is eerily prophetic for the nation of Kiribati in the twenty-first century. Climate change, caused by human actions that are out of their peoples' control and half the Earth away, is, among other global shifts, melting the ice at the poles and raising ocean temperatures, which has in turn raised average sea levels throughout the Pacific Ocean, even along the equator. In this way, the islands of Kiribati are slowly being over-washed, and their fragile freshwater sources, which are underground lenses, are spoiling due to the infusion of salt. Perhaps in our lifetimes the citizens of Kiribati, like the mother frigatebird, will no longer have any land at all on which their people can survive.

"The Song of the Frigate Te Itei" is not only a crucial story that features a symbolic mother seabird; it also teaches Western audiences that this animal in human stories represents far more than a dark pirate-like seabird or just a fierce aerial athlete, which is almost exclusively how frigatebirds have been depicted by Western sailors, naturalists, and writers.

For example, on September 29, 1492, Christopher Columbus and his fellow mariners were sailing on their first voyage across the Atlantic Ocean. "This morning I saw a frigatebird, which makes terns vomit what they have and then catches it in mid-air," Columbus wrote in his journal. "The frigatebird lives on nothing else, and even though it is a sea bird, it does not alight on the water and never is found more than 60 miles from land." (In the same entry Columbus describes terns catching flying fish that are about one foot long, about 30 cm, with "two little wings like a bat.")

As Columbus wrote accurately, frigatebirds do often force

Great frigatebird (*Fregata minor*): breeding male with inflated gular pouch, soaring over Pacific atoll.

smaller seabirds, like boobies or terns, to cough up the fish they've just caught. Biologists refer to the animal behavior of stealing prey from another species as kleptoparasitism. Frigatebirds catch the loot out of the air before it hits the sea. This is why Europeans had named these birds after types of agile naval ships, known as frigates. They are also known to sailors as man-o'-war birds, another type of naval ship.

The deeply forked tails and thin wings enable frigatebirds to turn sharply and accelerate in the air. They rarely land on the surface. In fact, they can barely swim or even walk. Like an agile, sharp-turning hang glider, frigatebirds have evolved to be exclusively creatures of the air, sometimes riding thermals and floating on trade winds far higher than most other birds, aquatic or terrestrial. Frigatebirds have the largest wing area to body weight ratio of any bird on Earth—even larger than that of albatrosses. A frigatebird can have the wingspan of a bald eagle, yet only bear a quarter of the eagle's weight.

Ornithologists today recognize five species of frigatebird. The species of which Columbus was familiar in the Cape Verde Islands and in the Caribbean is the magnificent frigatebird (*Fregata magnificens*), the largest of the five, with an up to eight-foot (2.4 m) wingspan. The frigatebirds that nest most commonly on the islands of Kiribati are the great frigatebird (*Fregata minor*), known in Kiribati as *Te etei are e bubara*, and the lesser frigatebird (*Fregata ariel*), named *Te etei are e aki rangi ni bubura*, of which there have been, by a recent count, over twenty thousand breeding pairs living in the Phoenix Islands group. The males of all five species have red gular pouches that they inflate like a balloon under their beaks to attract mates.

Mother frigatebirds lay only one smooth, white egg at a time. The males bring the materials that the females use to construct nests in low palm or mangrove trees or on top of other foliage. Their little chicks, fluffy with white down, are slow to fledge, taking five to six months to leave the nest completely. The fledglings remain dependent on their parents for food for a few months more. Breeding every other year or so and living at the very most for about three decades, magnificent frigatebird mothers might in the best of cases be able during their lifetimes to raise six or seven young to adulthood.

As observed by Polynesian and Melanesian navigators, and then by later sailors such as Columbus, frigatebirds do tend to stay coastal, which is useful for mariners trying to find land. But frigatebirds also fly over the open ocean far from their roosts. Some early Pacific voyagers likely kept frigatebirds on board their voyaging *waka*, their ocean-going boats, releasing the birds to reveal the direction of land. If too far from an island, the frigatebird would return to the boat.

Frigatebirds have long beaks with a hooked tip, like their cormorant and pelican cousins, which helps them to nab fish or squids from the surface. They've been observed following schools

Lesser frigatebird (*Fregata ariel*): mother and chick.

of dolphinfish to prey on the flying fish they scare up out of the water. Frigatebirds will also eat baby turtles and jellyfish, and chicks of other birds. Whalers in the Pacific in the 1800s reported that frigatebirds occasionally pecked at the flags on their mastheads, and mariners today have reported frigatebirds snapping at radio antennae at the masthead of their ships.

In the 1800s some of America's most famous writers had a special place for frigatebirds, but they mainly described their awesome flight and compared the birds to pirates. Artist-naturalist John James Audubon wrote in the 1830s that frigatebirds have "a power of flight which I conceive superior to that of perhaps any other bird." He described them diving for fish "from on high with the velocity of a meteor."

In 1876 Walt Whitman published the poem "To the Man-of-War-Bird," in which he describes seeing one from the deck of a ship:

Thou born to match the gale, (thou art all wings,)
To cope with heaven and earth and sea and hurricane,
Thou ship of air that never furl'st thy sails,
Days, even weeks untired and onward, through spaces, realms gyrating,
At dusk that look'st on Senegal, at morn America.

Perhaps no sea writer in English, however, gave the frigatebirds a greater role than Herman Melville. In his first autobiographical novel, *Typee* (1846), Melville describes seeing frigatebirds from a ship in the South Pacific: "That piratical-looking fellow, appropriately named the man-of-war's hawk, with his blood-red bill and raven plumage, would come sweeping round us in gradually diminishing circles, till you could distinctly mark the strange flashings of his eye; and then, as if satisfied with his observation, would sail up into the air and disappear from the view." Frigatebirds do have black plumage but not red bills—their beaks are usually grayish or tan. To be fair to Melville, though, some descriptions and illustrations of the mid-1800s did give frigatebirds red beaks, adding to their criminal aura.

In his novel *Moby-Dick* (1851), Melville wrote that a frigatebird, which he called a "sky-hawk" and a "sea-hawk," both of which were sailor synonyms for the frigatebird at the time, was pecking at a red flag that was at the top of the whaleship *Pequod* as it was sinking at the very end of the novel. After the ship goes down, taking the poor frigatebird, the "bird of heaven" under the water with it, another frigatebird swoops over Ishmael, the one survivor. In the story, Ishmael is now a floating orphan on a coffin made for Queequeg, the Pacific Islander hero, who had drowned along with the rest. Ishmael's fate is in the equatorial Pacific, arguably in the waters of Kiribati. Thus the frigatebird at the end of *Moby-Dick* is in its own way prophetic about human-induced climate change and the threats to the people of the islands of Kiribati, who, like so many Ishmaels, might become motherless orphans, might lose their homes to the seas, even as they are powerless against so many distant Ahabs in oil-burning nations on the distant edges of their Pacific Ocean.

"The Song of the Frigate Te Itei" is verse that is the perhaps the most meaningful description of a frigatebird today, inspiring those of us in industrial countries to work together, to rise up to slow global warming so that the mother frigatebird of Kiribati has somewhere safe to land and feed her children.

Grampus

If you happened to have been flipping through the pages of the Sunday *New York Times* on August 23, 1896, you would have seen on the second page, directly beside the announcement that Viceroy Li Hung Chang from Peking was making an official visit to Chinatown, this other headline:

FOUR GIRLS ON A SCHOONER
Made a Long Voyage and Now Spin Wonderful Yarns

The reporter explained that four young women had just returned to Manhattan from nearly a month's voyage, sailing round-trip from New York City to Nova Scotia aboard a merchant ship named the *Gypsum Empress*. The women, whose last names were Carew, Roach, and Clift (from Brooklyn) and Simpson (from Staten Island), had volunteered to travel on the Atlantic as working sailors. As the schooner carried its cargo of gypsum—a mineral for fertilizer, drywall, and chalk—the four friends learned the ropes, stood their watches, rode through a fierce squall that splintered a wooden boom, and just had a fabulous time overall. They each kept a journal, sketched, and led their entries with quotations from the famous ocean poet Henry Wadsworth Longfellow.

"The most exasperating calms and dispiriting fogs marked the early part of the homeward voyage," the reporter continued, "but a grampus, or cowfish, some twenty feet long enabled Miss Roach to bring her Kodak into play."

A grampus? A cowfish?

In the 1800s, the words *grampus* and *cowfish* were pretty fluid. The word *grampus* probably comes from the simple joining of two French words: big (*grand*) and fish (*poisson*). By the time of this voyage on the *Gypsum Empress* in the 1890s, at least in the US fisheries reports, both *grampus* and *cowfish* referred to the Risso's dolphin (*Grampus griseus*), terminology which has continued up through today.

But for English-speaking mariners earlier in the nineteenth century and maybe even on the *Gypsum Empress*, references to a grampus or cowfish could have meant a pilot whale (*Globicephala* spp., the "Blackfish" drawn by Thomas Albro), a killer whale (*Orcinus orca*), or any of a few other large ocean dolphins with blunt snouts and large dorsal fins that curl backward like a crescent.

Risso's dolphins, pilot whales, beaked whales, killer whales, and a few other large ocean dolphins are all found in the Gulf of Maine in the summer. It can be pretty hard to estimate the size of dolphins and whales at sea ("dolphins" being, somewhat arbi-

Risso's dolphin (*Grampus griseus*) with body scratches and ventral crease in the melon, or forehead.

trarily, just smaller toothed whales) and to tell some of these species apart—especially when you're leaning over the side of the ship (and trying to take a photograph with an old-timey Kodak). But if their grampus stuck around for a while and swam quite close, they would have been able to rule out a killer whale, since killer whales have such distinctive white patches and exceptionally tall dorsal fins. Risso's dolphins are generally paler gray than these other species when they grow older, and Risso's dolphins are also more often scratched with scrapes and scars from the teeth of others of their own species, as well as from skirmishes with squids and maybe sharks. So if their grampuses were indeed Risso's dolphins, the four women in true sailor fashion likely exaggerated the twenty-foot (6 m) length, since Risso's dolphins rarely get larger than thirteen (4 m) feet, male or female.

As for *cowfish*, the nickname seems to have been ascribed to a few ocean dolphin species over the centuries, but I personally don't see much likeness. It makes a bit more sense to me to apply the name to the large, vegetarian, walrus-like manatees, the "sea cows."

Risso's dolphins, like belugas and other ocean dolphins, have melon-like heads that contain special fats and oils to assist with echolocation. Risso's dolphins swim down into the dark as far as 1,000 feet (305 m) to hunt fish and squids, sometimes for as long as a half-hour on a dive. Like all toothed whales of any size, these grampuses have a single blowhole for inhaling and exhaling, and Risso's dolphins have a distinctive vertical crease along their melon. They have conical teeth in their lower jaws for chewing their fish and squids, and they have fewer pairs of teeth than a lot of their cetacean cousins, with a slight underbite—which might look just a bit bovine.

Risso's dolphins are not just common in the Gulf of Maine. They swim in tropical and temperate oceans as far north as Alaska and the Shetland Islands and as far south as Tasmania and Cape Horn. Risso's dolphins are most commonly found in offshore waters, but occasionally they live quite close to the coastline.

In the 1890s, at the same time that these four young people were sailing across the Gulf of Maine, a particular grampus on the other side of the world, a Risso's dolphin, was gaining fame in Aotearoa New Zealand because it continually followed ships, leaping in their bow wakes within Cook Strait (the same frigid, tumultuous strait that Lynne Cox swam a century later). The sailors named this whale "Pelorus Jack," after Pelorus Sound near where it regularly met the ships. Pelorus Jack became, arguably, the first legally protected animal in the world when the Governor of New Zealand signed an official order in 1904 to protect this Risso's dolphin, because there had been reports of people trying to harpoon this region's mascot.

The English author Rudyard Kipling saw Pelorus Jack himself, the "big, white-marked dolphin," on his trip to Aotearoa New Zealand. Then in 1897, the year after those four greenhand sailors made the *New York Times*, Kipling published his novel *Captains Courageous*, in which a young wealthy boy named Harvey finds himself aboard a fishing schooner with another boy, Dan, who is already an experienced fisher on the Grand Banks. In one scene, just to the east of where the four women sailed home from Nova Scotia, Kipling writes of grampuses, pointing out their scratched-up whiteness:

> At the first splash a silvery-white ghost rose bolt upright from the oily water and sighed a weird whistling sigh. Harvey started back with a shout, but Dan only laughed. "Grampus," said he. "Beggin' fer fish-heads. They up-eend thet way when they're hungry. Breath on him like the doleful tombs, hain't he?" A horrible stench of decayed fish filled the air as the pillar of white sank, and the water bubbled oilily. "Hain't ye never seen a grampus up-eend before? You'll see 'em by hundreds 'fore ye're through."

When Carew, Roach, Clift, and Simpson returned from their journey aboard the *Gypsum Empress*, they reported, in addition to their grampus, a much larger type of whale. It's worth considering how excited they were to see these whales, the grampuses and the larger ones, showing that the thrill of seeing them from the rail of a ship and our eagerness to photograph whales at sea was not something that began in the 1960s.

The reporter wrote how Captain Roberts, who had known all four of them since they were little girls, reacted as the new sailors—wind-blown, tan, and still swaying a bit on the dock—

yarned about their trip: "[The captain] squirmed a little when they told about two waterspouts they had seen. That was near Cape Cod on Tuesday. They had just been taking photographs of three mammoth whales, at least sixty feet long, and disporting themselves 'hard by on the port hand,' as they all echoed at once. The whales were just too lovely for anything."

Green Turtle

Once there was a sea turtle caught by a fisher; it escaped by a lucky break and then had the miserable misfortune five months later to be caught a second time. The only silver lining was that this event helped build the case for sea turtle conservation around the world.

The animal was an exceptionally large green turtle (*Chelonia mydas*), perhaps over four feet long (1.2 m) and maybe over four hundred pounds (180 kg). Adult green turtles have rounded snouts and serrated beaks, evolved to munch seagrasses. Their diet is part of the reason why green turtles have been hunted so heavily by humans for thousands of years. Apparently the greenish-colored muscles, the meat of these vegetarian turtles, are tastier for people—as opposed to, say, the leatherback and hawksbill turtles, whose diet consists mostly of sponges, jellyfish, and crustaceans.

This particular sea turtle's life began almost certainly just as did that of all other sea turtles for tens of millions of years before him. His mother, with her rear flippers, dug a hole in a beach on a remote sandy shore of the western Caribbean. She deposited about one hundred eggs, covered over the hole with sand, wished them all *buena suerte* and *vaya con Dios*, and then she left. After about two months as an egg, the turtle hatched along with the one hundred or so others over the course of a couple days and climbed out of the hole. Each of the baby turtles was no bigger than a playing card. Somehow, through intuition and

Green turtle (*Chelonia mydas*) with attached remora and free-swimming pilot fish.

genetics, they all just knew, even though they could barely see, that they needed to make it to the water—and fast. Several of the sea turtle's siblings might have been too slow or just unlucky, eaten that night on the beach by a heron or a vulture, or even by a young jaguar. Once the rest made it into the water, several more baby turtles were likely eaten by underwater predators, such as jackfish or rays, while the survivors managed to swim to depth and some relative safety.

Sea turtles have been around in a similar form for over 140 million years, once swimming with the plesiosaurs in the Jurassic seas and now evolved today into seven separate species in all tropical and temperate waters. Among the over twelve thousand species and subspecies of reptiles on planet Earth, sea turtles are among only a few groups that have returned to the sea for their living, including the marine iguana of the Galápagos, the marine crocodiles, and about eighty species of sea snakes, including the yellow-bellied sea snake. Green turtles not only are different from the other sea turtle species in their diet of seagrasses and algae, but female green turtles are the slowest to reach sexual maturity, unable to reproduce until they are about 25–40 years old, depending on their diet and location around the world.

In the wild, green turtles might be able to live until sixty. It's not known exactly how old this particular twice-caught giant was, but at some point in his life he survived an attack by sharks or a school of tuna or dolphinfish, leaving all four of his flippers scalloped by a series of bites, scarred for life, rendering him immediately recognizable to fishers, like a reptilian Harry Potter.

The first capture was in 1924. Captain Charles Bush, a veteran skipper-owner of a turtle schooner out of the Cayman Islands, south of Cuba, had sailed all the way down to Mosquito Cay, just off the Nicaraguan coast. As was the custom, Captain Bush branded his initials into the green turtle's shell.

By the 1920s, Caymanian captains had been sailing down to the Central American coast for over a century, ever since their own waters had been fished out of turtles by early European mariners and colonists. Indigenous people, here and throughout the Caribbean, however, had been hunting green turtles since they first figured out that they could push them over on the beach. Once upside down, it's hard for a turtle to flip back, making them easy prey. Some Indigenous communities of the Caribbean also perfected a method to catch turtles out at sea using nets or harpoons. The Miskitu people of Central America, for example, became famous for their skills in boats and with spears in catching sea turtles. When the first Spanish and English explorers arrived in the Caribbean in the 1500s and 1600s, these mariners wrote of the Miskitu's skill at catching turtles and how abundant sea turtles were at the time.

The European surgeon Alexander Exquemelin wrote in 1678, "The Indians often go to sea with the rovers [pirates], and may spend three or four years away without visiting their homeland. . . . These Indians are a great asset to the rovers, as they are good harpoonists, extremely skillful in spearing turtles, manatee and fish." Exquemelin continued, "An Indian is capable of keeping a whole ship's company of 100 men supplied with food."

Pirates, in fact, kidnapped Miskitu turtle hunters to help supply food for their ship's crew. Sailors have always loved to eat turtle because, especially in the times before refrigeration, with ships struggling with scurvy, the nutritious animals could survive on board for weeks at a time, greatly advancing the ability of colonial mariners, for better or worse, to sail to all corners of the Caribbean. Historian Sharika Crawford explains, "The expansion of exploration and peopling of the region was dependent on sea turtles."

So, after Captain Bush caught the scarred green turtle in a net and branded the shell, he sailed it up along with other captured sea turtles to Key West, Florida, where he sold them to local merchants who placed the turtles in a fenced area in shallow water, known as a *kraal*, to keep the animals alive before shipping them elsewhere or killing them for turtle soup or canned turtle meat.

Yet before the large scarred green turtle was killed, an October hurricane blew through and wrecked the kraal. All the turtles escaped.

This is where the story gets truly exceptional. During the next season, Captain Bush was again fishing off Mosquito Cay. Five months after he'd caught it the first time, the fisher identified the turtle's size and scalloped flippers. He caught this turtle again in his nets, found his brand, sailed him up to Key West with the others, and sold the turtle once more. The merchants there were none the wiser, unaware that one turtle in the kraal was worth two for Captain Bush.

Some years later, along came a marine biologist named Archie Carr, who would go on to be an important advocate for sea turtle conservation around the world. Carr understood that he had to learn from and serve the Indigenous and Caymanian communities of people in the region if he wanted to care for sea turtles, too.

"The most serious handicap in any effort to save the green turtle is our ignorance of its migratory movements," Carr wrote in 1954. "Fishermen everywhere believe the green turtle migrates. Such a belief can also be found in the writings of the naturalists. But nowhere in the canons of zoology is there a shred of what could be called scientific evidence to prove it."

So Carr talked to the Caymanian turtle hunters and heard, among several other stories, this one about Captain Bush and the twice-caught turtle. Carr designed tagging studies to trace green turtle migration as he began to teach Americans and Europeans about the turtles' extraordinary passages and power of navigation, thus working to help protect the nesting beaches of sea turtles in the Caribbean and all around the world.

Scientists today theorize that sea turtles find their way by using a combination of magnetism, wave direction, and sensitivity to solar and lunar light—and maybe also sound and smell. It's about nine hundred miles (1,450 km) from Key West to the islands off Nicaragua.

According to the International Union for Conservation of Nature (IUCN), all species of sea turtle for which there is acceptable data are now classified as either vulnerable (loggerhead, leatherback, and Olive's ridley), endangered (green), or critically endangered (hawksbill and Kemp's ridley). Many sea turtle populations seem to be recovering in the Caribbean region, but too slowly, and the global populations of all sea turtles are declining. In most countries it is now illegal to export or import sea turtles. Scientists estimate that before European colonialism, tens of millions of green sea turtles, quite literally, ranged throughout the Caribbean. Swimming along their millennia-old migratory highways, green turtles are now only at about 1 percent of their historical population. And as they fly underwater through the Caribbean Sea, their paths from feeding grounds to nesting beaches, even with the use of DNA and other advanced research methods, remain mostly a mystery to humans.

Guanay Cormorant

Elisa Goya Sueyoshi wears light clothes that cover her head, face, and the rest of her body to protect against the sun and the dust. She's even tucked the bottom of her pants into her socks to protect against the leaping ticks, known in Spanish as *garrapatas*.

Followed by her two research assistants, Goya Sueyoshi stoops low and searches in a narrow lane between nests. She's looking around on a cliff of a bare, brown island named Isla Chincha Centro off the coast of Peru. She and her assistants are looking for cormorant pellets, which are similar to the pellets that owls hack up, except those from cormorants are not made of undigestible rodent fur and bones; these are walnut-sized, soft, boogery balls filled with fish bones and bits of crustacean shells.

Nearly all of the cormorants have flown off at this point, but Goya Sueyoshi still doesn't want to take too much time here in case there might be a stray nest with eggs or one with young who might overheat in the open sun or be taken by the gulls.

Goya Sueyoshi spots a pellet just beside a nest to her left. She slips it into a plastic bag. Then a few minutes later, she smiles and shows her assistants an even better prize: a regurgitant. This is even more valuable than hacked up pellets, because it is a true pile of vomit, a fist-sized mass, still warm and filled with identifiable, partially digested whole fish, including at least two anchovetas (*Engraulis ringens*). One of the young cormorants had just spat this up in panic, before it flew away. Goya Sueyoshi lovingly scoops this regurgitant into another plastic bag.

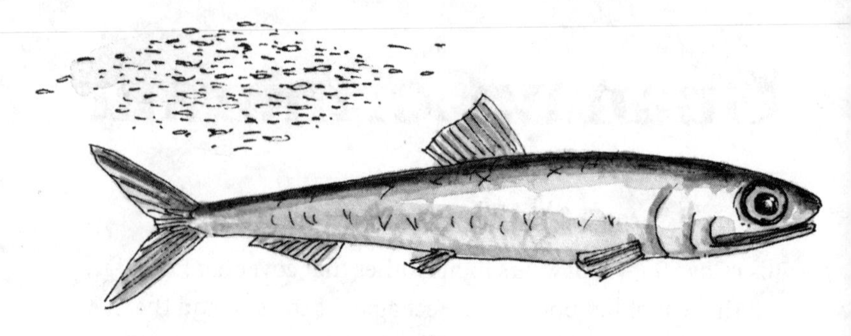

Peruvian anchoveta (*Engraulis ringens*).

Dr. Elisa Goya Sueyoshi is a marine biologist. It is January, 2002. She works for the Peruvian government agency that monitors fisheries populations. The seabirds on Isla Chincha Centro are intimately connected with the ecology of the schooling fish, the anchovetas of the mighty Peru Current, which by the tens of millions stream northbound past these islands. Goya is here collecting cormorant pellets today because the thousands of seabirds on this island eat fish that humans want to catch. She is also here because the cormorants that eat these fish produce prodigious amounts of white liquid poop. And because of their guano, this bird species, aptly named the guanay cormorant (*Leucocarbo bougainvillii*), was declared in 1925 by the American ornithologist Robert Cushman Murphy to be "the most valuable bird in the world."

The guanay cormorant is one of about thirty-two different cormorant species that live and hunt fish beside nearly every large body of water on Earth. The more seagoing, more slender of the cormorants are often called "shags" by English-speakers. Unlike nearly any other animal on the planet, cormorants regularly perform an ultimate triathlon: they dive daily over a hundred feet below the surface, waddle around on land, and not only do they fly around each day, but several cormorant species migrate thousands of miles each year. (Consider, for example, that chinstrap penguins can walk and swim, but they cannot fly; while frigatebirds are champion fliers, but they cannot swim underneath the surface at all and can barely walk more than a few steps ashore.) Humans have yet to invent a single automated object that can do all that one cormorant can. Cormorants live beside fresh and saltwater, at sea level, and at elevation. Cormorants poke around reefs in coastal waters along the equator and dive beneath icebergs within the Arctic and Antarctic Circles.

Cormorants are colonial birds that fish during the day. They are famous for spreading their wings out to dry, Dracula-style, although not every cormorant species exhibits this behavior. Biologists believe they need to do so because their feathers evolved to better absorb water, so that when they are diving deep for fish, this more wettable plumage acts like the weight belt used by scuba divers or the haenyeo. Once out of the water, though, cormorants often need to dry their feathers. At night, cormorants fly back to their home to roost and to nest together tightly for safety. Cormorants, like most seabirds, prefer to live on islands so as to be safe from mammals that would eat their eggs and young.

Guanay cormorants, which live only along the western coast of South America from Ecuador to Chile, have mostly glossy black-green plumage with a white breast and neck. Guanays have pink feet, a wide ring of red skin around each eye, and both males and females, when breeding, have a jaunty little faux-hawk crest of feathers on their head. Guanay cormorants live in some of the densest colonies of any other cormorants—of any seabirds, in fact—cramming themselves together at times with three or four nests within a single square yard (.8 sq. m). That's two adults per nest, each with a wingspan of over two feet (.6 m). During breeding season, they cram in an additional two to four chicks. Of the

Guanay cormorants (*Leucocarbo bougainvillii*) on the Chincha Islands, Peru.

guanay cormorants, expert Bryan Nelson declared: "A more gregarious bird is beyond imagination."

Historically, there were likely over one million cormorant nests on Isla Chincha Centro alone, which is within swimming distance of two similar islands, Norte and Sur, and within a string of a dozen or so other coastal guano islands dotted along the coast that were once also crammed with colonies of seabirds. These islands are not only uniquely situated within this enormous fishy food source, but also, because of the oceanography and geology of the coast, the islands are nearly entirely free of rain, allowing centuries upon centuries of seabird guano to dry and accumulate in some places reportedly over three hundred feet (90 m) high.

In the millennia before European contact, the Pre-Incan and Incan peoples along the west coast of South America first discovered the extraordinary fertilizing capabilities of seabird guano.

The three Chinchas are among the islands closest to the shoreline. Archaeological studies and historical records reveal that people paddled or sailed out to harvest the guano, returning to pack it in baskets or bags, and using llamas to haul these up into the hills to spread on their fields of corn and other crops. When Spanish people arrived in the 1500s and 1600s, they learned to fertilize their own crops with guano at their early settlements here. Peruvian boobies (*Sula variegate*) and Peruvian pelicans (*Pelecanus thagus*) also nested by the hundreds of thousands on these islands, but it was the guanay cormorants that were historically by far the most populous and thus the most prolific providers of poop. In the early 1800s, Alexander von Humboldt and his botanist colleague Aimé Bonpland brought guano samples back to Europe. It wasn't long until there was a full-blown "guano rush" for this miracle fertilizer. For example, over some months in 1841, twenty-three ships pulled alongside the docks of Liverpool, their holds filled with a total of over sixty-three hundred tons of bags filled with dry, noxious cormorant feces.

The guano rush coincided with the gold rush in northern California, so after dropping a load of people off in San Francisco, captains of some of the largest, fanciest clipper ships anchored on their way home at the base of the cliffs of the Chincha Islands to fill their holds with bags of dried, aged cormorant droppings. Harvesting the guano, shoveling this toxic dust into bags, was brutal and dangerous work. Indigenous people quickly grew ill and exhausted working for colonial overseers, so merchants replaced them by convincing indentured servants from China, effectively slaves who were unaware of the horrors that awaited them, to sail across the Pacific and work on the islands. Some merchants even raided Pacific Islands to kidnap laborers to shovel cormorant guano.

Within a couple of decades, however, the guano rush was

over. The harvest had been so aggressive, the islands and seabirds so disturbed, that the guano resources were nearly depleted. To meet demand, people found cheaper fertilizers, like rock nitrates, and then soon developed lab-produced chemical fertilizers.

In the early 1900s, local Peruvian and Chilean officials began to revive the guano trade for a more local agricultural market. This began with protecting the seabirds and making sure their nesting areas were safe from mammals and poachers. Managers and scientists were pleasantly surprised to find that the guanay cormorant and other bird populations came back quickly. Instead of centuries, it really only took decades for the birds to deposit harvestable swaths of guano. But then in the 1950s and '60s, fishers began to overharvest the anchovetas in the Peru Current. This devasted the recovering cormorant populations, which were, among other factors, perhaps more dependent than the boobies and other seabirds on those fish.

By 2002, when Elisa Goya Sueyoshi and her team were collecting pellets, both fisheries and seabird populations were slowly on the mend, with over a half-million Peruvian boobies living on Chincha Centro, but only about forty-five thousand guanay cormorants. Goya Sueyoshi and her colleagues collected the pellets and regurgitants that January day to determine the seabirds' diet. This is a way to monitor bird populations as well as those of the anchovetas, both of which have for millennia fluctuated based on El Niño phases in the Pacific—long before humans began destroying the seabirds' island habitat or netting up millions of anchovetas.

Today, as they have since the early 1900s, local people, *guardias*, live on the island year-round to protect the birds and the guano, since there is still an organic fertilizer market in Peru and overseas. And Goya Sueyoshi, two decades later, still visits the Islas Chinchas with research assistants to collect pellets and regurgitants and to count the birds. The total numbers of guanay cormorants on the Chinchas, however, as well as on the rest of the islands along the coast continues to decline.

Halibut

The painter Winslow Homer is standing in his seaside studio in Prouts Neck, Maine. He holds in one hand his palette, dabbed with greens and grays and blues. He pauses occasionally to put his brush down, to step back, to look at the painting, and likely to twirl the waxy tips of his handlebar mustache. It is 1885.

On the canvas, which is four feet long by two and a half feet high (1.2 × .8 m), a lone fisher stares over his left shoulder at the horizon and his schooner. The man in the painting is looking at a dangerous fog bank approaching, which he knows will soon envelop the horizon and make it more difficult, if not impossible, to row over the steep seas and find his ship. Homer depicts the quiet heroism of this fisher: a handsome profile, the man's strong, calm jaw tilted upward, his sou'wester rainhat looking like some sort of Viking helmet.

The summer before, in order to gather material for this painting and several others about North Atlantic fishers, Winslow Homer traveled to Gloucester, Massachusetts, a place he knew well. Here he sketched boats and people, and even went to sea aboard fishing schooners. In Gloucester he surely heard about and perhaps even met Howard Blackburn, a man who had recently been separated from his schooner in a January blizzard. His dorymate eventually froze to death, while Blackburn, his bare hands frozen to the oars, managed after five days at sea to row the boat and the body of his dead friend to a desolate cove in Newfoundland. Blackburn lost all of his fingers and half of his toes to frostbite. Homer would have learned in Gloucester that 209 men had died out sea while fishing during the previous year.

After his research in Gloucester, Homer returned to Prouts Neck and sketched out his painting. He hired his handyman, Henry, to model for him by sitting in a dory propped up in the sand. He splashed Henry with a bucket of seawater to get the proper effect on the oilskin coat, but apparently this was not part of the arrangement, and Henry let his boss know it with a volley of curses.

As *The Fog Warning* takes shape, Homer brushes bright whites onto the underbodies of two enormous fish plopped in the stern of the boat. The fish are front and center in the painting, their bellies pale and glistening. He paints his fisher's boat tilting upward on a wave, so the inside of the boat and the fish are highlighted, drawing the viewer's eye, the white patches of the fishes' underbodies parallel to the crests of waves that slosh dangerously close.

Homer's fish here are halibut, the same fish that Blackburn and his dorymate were trying to hook on that fateful winter day.

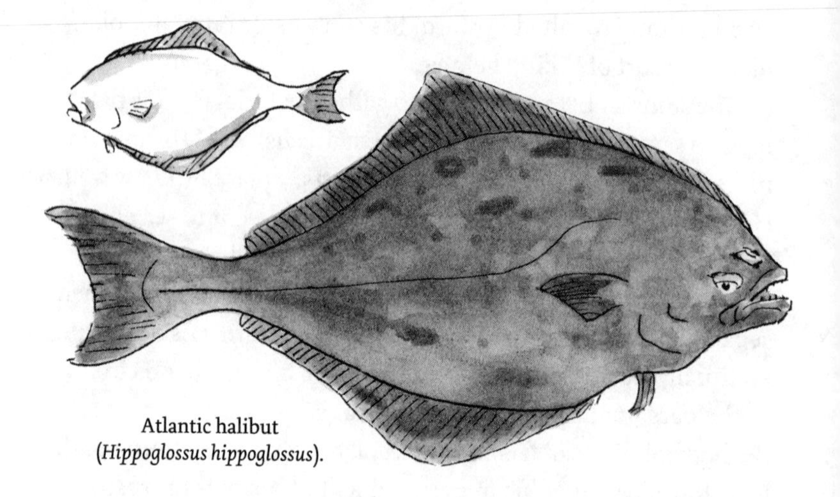

Atlantic halibut (*Hippoglossus hippoglossus*).

Homer had originally wanted to title the painting *Halibut Fishing*, but someone, thankfully, convinced him this lacked the drama of *The Fog Warning*.

Halibut, of which there are three species, all in the northern hemisphere, are the largest of the world's flatfish, a broad group that includes flounders, soles, and turbots. Halibut appear to swim sideways, because both eyes are on one side, the "right" side of their head, so the other side can lie flat on the sea floor while the animal can still look up and around. Its right side, the side with the eyes, is olive green, gray, or chocolate brown, and often splotchy, particularly when it's younger, so they blend into the sand, gravel, or clay of the ocean bottom. The halibut's "left," or eyeless side, is white or speckled gray. Halibut have sharp curved teeth to eat other fish, lobsters, quahogs, or even a seabird if one happens to be dozing on the surface.

Atlantic halibut (*Hippoglossus hippoglossus*), the species in *The Fog Warning*, were once plentiful and large throughout the western North Atlantic. As early as 1637, English colonist Thomas

Morton reported, "There is a large sized fish called Hallibut, or Turbut: some are taken so bigg that two men have much a doe to [haul] them into the boate; but there is such plenty, that the fisher men onely eate the heads & finnes, and throw away the bodies."

Fishers arriving from Europe primarily caught Atlantic halibut on the Grand Banks, Georges Bank, and in the coastal waters of the Canadian Maritimes, but these fish range all the way around the North Atlantic, from Virginia north up to Iceland and the Barents Sea and around south to the Bay of Biscay. For centuries North Atlantic fishers caught enormous numbers of these halibut, but usually, as Morton explained, colonists considered this species secondary to cod, haddock, and other large, edible fish, mostly because halibut were usually too big and thick for salting. This was before fishing boats had the ability to carry ice, so halibut would go bad before they could get them to market. The men usually threw halibut back into the water or just cut off their heads and fins, tossing the rest into the ocean. Even in the early 1800s, according to fisheries experts at the time, halibut were easily caught—even close to shore in Massachusetts Bay and off Cape Cod, "so much so as to be sometimes regarded as a decided nuisance by cod fishermen."

Atlantic halibut average 50–150 pounds (23–68 kg). Females grow much larger, and several reasonably trustworthy accounts report individuals hauled up on a boat's deck or into a dory that weighed over 200–300 pounds (90–135 kg), including one in 1917 that weighed about 700 pounds (~318 kg) and measured over nine feet long (2.7 m). Biologists believe individual halibut can live for fifty years.

Previously considered nearly worthless and a nuisance, in the 1840s the demand for Atlantic halibut began to expand, in part because of the new ability to ice the catch and to move the prod-

uct by train. America also saw a general shift in tastes toward eating more fresh fish, especially influenced by European immigrants, who ate more fish. Around this time, too, since it was no longer effective enough to just fish over the rail of a ship, schooners began the system of sending out dories, the small wooden boats, usually with two men in each. Thus, just as Winslow Homer depicts, fishers rowed out from a schooner and put out long lines with hooks over the sides, usually baited with smaller fish or even the meat from murres, cormorants, and other seabirds killed for this purpose. Within only forty years, however, New Englanders and Canadians overfished Atlantic halibut until they were difficult to find at all: the fishery fell flat.

By the mid-1880s, as Winslow Homer was brushing his finishing touches on *The Fog Warning*, some men from Gloucester were already sailing all the way to Greenland or Iceland to try to bring home a load of full-sized halibut. Fishers were looking over their shoulders at clear signs of the end of commercial fishing for this species, which today gives the word *warning* in Winslow's title an additional meaning. In the decades to come, fishers out of Gloucester and other ports had to set more and more hooks and launch more and more dories to catch other species, such as cod and pollock, until the development of otter trawls and the use of marine engines in the 1920s. Notably, Atlantic halibut, large fish that take several years to mature, were harvested to near extinction in the western North Atlantic *before* any of these more-industrial inventions.

Today, Atlantic halibut are still a long way from getting back to the populations observed in the days of Thomas Morton's 1600s, even though halibut fisheries in the North Atlantic and in the North Pacific are carefully managed. Conservation groups do suggest, if you eat fish, that wild-caught halibut is actually a pretty good choice these days—if you can afford it.

And now you can turn to a friend at the Museum of Fine Art in Boston and ask why Winslow Homer dabbed such bright whites on the huge, flat fish centered in his masterpiece *The Fog Warning*. When your friend says "Why?" you can properly quip, "Just for the halibut."

Horse

James H. Williams grew up in Rhode Island. His loving father, a well-educated sailor and a Long Island Sound pilot, died when James was not yet seven. But it had always been clear that he had wanted his son to go to college. James had curly, reddish hair and light-brown skin. He fell in love with the sea as a child, so his mother eventually allowed him to apprentice to a merchant ship captain at twelve years old. When the ship was in Barbados, his captain, his legal guardian at that time, died. This left James alone at barely fourteen working aboard a ship among strangers in the Caribbean. So began his decades as a merchant ship sailor and a whaler. Over time he proved himself a gifted writer about life at sea, which included an important record of a festival on ships involving a poor old horse.

This story began in January of 1888, when Williams, now twenty-three years old, was in Philadelphia. Here he signed onto a ship bound for Calcutta. Just before he left, he learned of the death of his mother. She was white and had married his African American father at a time when these marriages were illegal. Despite whatever racism they encountered, James wrote that his childhood had been overall a happy one. The loss of his mother was especially crushing to him, even though he had spent so much time at sea.

His new captain and the men at the shipping office refused to grant Williams shore leave to go to his mother's funeral. They claimed the letter was forged and believed he was planning to escape with the cash advance they had given him for signing on as crew.

Shipping agents at the time gave crewmembers advance money when they committed to work aboard a given ship. The men could then leave the advance with loved ones ashore, pay off bills or debts in port, or they could use the money to buy clothing and supplies for the upcoming voyage. When Williams protested that he was genuinely not trying to escape with his advance, that he really did need to go to the funeral, they called the police. Williams was beaten with clubs, thrown in jail, charged with desertion, then forced back aboard.

"So I sailed away with a heavy heart," Williams wrote, having never been able to pay proper respects to his mother.

After two months out at sea on this ship, Williams explained that his "dead horse" was now worked up. For decades, this had been a tradition aboard ships after the crew had started to make money for themselves beyond the advance they owed to the ship for signing. For this festival, the sailors hauled up a horse to the yardarm while singing a specific chantey, a work song for pulling together, with the refrain: "Oh, poor Old Man, your horse must die! And I think so; and I hope so!"

Williams wrote that at the end of the song they cut the halyard, and the old horse dropped into the ocean.

This was not a real horse, but an effigy, a large puppet, which the crew likely made with old sail canvas, worn rope, maybe a spare barrel for the body and pieces of scrap wood for the legs. The dead horse festival marked when that advance, the debt to the ship was now paid off, and the sailors would begin making money for themselves again. On some ships this was a day-long celebration, a theatrical event in which the horse puppet was paraded around, auctioned, and after they sang the chantey they raised the puppet aloft where they even sometimes lit the thing on fire or whizzed off a flare before cutting the line and dropping the "poor old horse" into the sea. Sailors have always enjoyed most any holiday or festival to break up the monotony at sea or to balance difficult times. The dead horse ceremony was a rare something to enjoy and celebrate.

During James Williams's decades on the ocean, he probably saw a few different versions of this dead horse festival. He also would have known a half-dozen or so other references to horses on merchant ships in the 1800s, none of which are all that favorable if you love horses. The sheer number of these horse references (as well as those to pigs, cows, and chickens) shows how often sailors brought vocabulary from farm life ashore to their shipboard language. For example, an iron bar on deck, used to shift a sail from one side to the other, was called a "horse." The footrope that Williams and his fellow sailors used to stand on and balance themselves when working to furl a sail on a yard was also called a "horse." White crests of foam on the tops of waves are still called "horses" by mariners, and a type of particularly wispy cirrus cloud is still known at sea as a "mare's tail." When the preserved beef was really tough in the 1800s, the sailors would call it "salt horse." (It's possible even occasionally there was indeed some actual horse meat fed to sailors, although food historians believe that was rare). And beyond the "Poor Old Horse" chantey, there's another song dating back at least to the 1820s, still alive among some of today's sailors, with the line: "Sailors at sea earn their money like horses, To squander it idly like asses ashore." This has become a mantra among some twenty-first century sailors—"hosses at sea, asses ashore"—equivalent to "work hard, play hard."

A final example for the use of the word *horse* at sea is the "horse latitudes," a region of the doldrums, in which there can

be periods of long calms at sea at about 30° North latitude and about 30° South latitude in both the Atlantic and Pacific Oceans. Some claim dubiously that the derivation for this phrase is that horses were regularly thrown overboard at these latitudes, either because the animals died of thirst or even that sailors sacrificed the horses to reduce the ship's weight so that it would sail better in the light air. Some historians also contend that this was about the distance from England or the American northeast in which sailors would now have worked off their advance and have these dead horse ceremonies in these "horse" latitudes. The truth is that we should rein in these derivations. We really don't know for sure where the name for the "horse latitudes" came from, or when or where or why the dead horse festival began—or even exactly why it's a poor old horse puppet as opposed to any other symbol to celebrate the sailors' earnings.

For his part, however, James H. Williams left historians one of the best descriptions of the dead horse ceremony from a sailor's perspective. Williams wrote dozens of other articles, too, especially for an influential magazine at the time called *The Independent*, in which he advocated for unions, new laws, and better wages to improve the lives and treatment of sailors.

Williams died in 1927 after his retirement at Sailor's Snug Harbor on Staten Island, New York, which was then a state-sponsored retirement home for aging merchant seamen. Sailor's Snug Harbor had dining halls, gardens, farms, and a bakery. James H. Williams's essays of life at sea and its hardships, which he continued to write into the 1920s, were compiled into a manuscript that he continued to work on, published after he died as *Blow the Man Down: A Yankee Seaman's Adventures Under Sail* (1959). I like to imagine his final days at Sailor's Snug Harbor as peaceful ones, perhaps even leading a tired old horse along the path beside New York Harbor.

Isurus oxyrinchus

A mako shark (*Isurus oxyrinchus*) shoots up from the deep sea and bursts through the surface so high and powerfully that its dark blue back, like a tuna's, glitters in the sunlight. The enormous animal splashes back down, thuds into the water, then settles to swim directly toward the scent of blood and fish oil. Its high stiff dorsal fin slices through the surface.

This scene is from Ernest Hemingway's novella *The Old Man and the Sea* (1952). The author writes, "He was a very big Mako shark built to swim fast as the fastest fish in the sea and everything about him was beautiful except his jaws."

Hemingway was living in Cuba when he wrote the story about his protagonist Santiago, an aging Cuban fisher. In the Spanish translation of *The Old Man and the Sea*, Lino Novas Calvo translates the description of the shark's entrance this way: "*Era un tiburón Mako muy grande, hecho para nadar tan rápidamente como el más rápido pez en el mar, y todo en él era hermoso, salvo sus mandíbulas.*"

"Inside the closed double lip of his jaws," continues Hemingway, "all of his eight rows of teeth were slanted inwards. They were not the ordinary, pyramid-shaped teeth of most sharks. They were shaped like a man's fingers when they are crisped like claws. They were nearly as long as the fingers of the old man and they had razor-sharp cutting edges on both sides. This was a fish built to feed on all the fishes in the sea that were so fast and strong and well armed that they had no other enemy."

The old fisher, Santiago, knows the mako shark as *dentuso*, meaning 'big-toothed.' He is a poor, once proud person—once a turtler alongside the likes of Captain Bush—who one morning, after failing to catch fish for months, decides to try going much farther into the Straits of Florida, farther than any of the other fishers from his village. Using bait caught on the way out and a hand line, Santiago manages to hook a monstrous marlin. After three days of struggle and superhuman endurance, the old man kills the marlin with his harpoon. At over fifteen hundred pounds (680 kg), it is so large that he must lash it to the side of his boat and try to sail the prize home.

The old man knows the sharks will come. And los tiburónes do come—with what feels like vengeance.

Ernest Hemingway is now better known for his stories and exploits with big game hunting and bullfighting, yet it was his *The Old Man and the Sea* at the end of his career that earned him the Pulitzer and Nobel prizes. For nearly thirty years, Hemingway was a devoted deep-sea sportfisher and often spent more than half his year on the waters of the Straits of Florida. In 1934 Hemingway bought his own fishing vessel, *Pilar*, which he piloted regularly into the Gulf Stream from Key West, Florida, and then later from his home in Cojímar, Cuba. From aboard *Pilar*, Hemingway fished alongside his hired crew. During the Second World War, Hemingway volunteered to search for enemy submarines, looking for the same U-Boats that torpedoed Poon Lim's ship. Hemingway meanwhile helped the observations and collections of marine biologists. Onto the shelves of *Pilar*, he crammed field guides on fish and seabirds alongside texts about oceanography and marine biology. Among other projects, Hemingway helped catch and observe sharks for biologists in Florida, New York, and Woods Hole. He took shark specimens for wartime researchers who were studying shark repellents. As a sport fisher, Hemingway claimed to have nearly caught a 1,000 pound (455 kg) mako off Key West, and he held for a time the North Atlantic world record for catching a 786-pound (356.5 kg) mako with a rod and reel in 1936. In short, when Ernest Hemingway wrote of a Cuban fisher bobbing around in the middle of the Straits of Florida in *The Old Man and the Sea*, he knew what he was talking about, especially when it came to *Isurus oxyrinchus*.

In Hemingway's story, the mako shark smells the blood and oils of the marlin. It attacks the dead fish beside Santiago's boat, tearing off chunks and spreading still more blood and chum into the water. The mako, the *dentuso*, does have unique teeth as compared to other large sharks, evolved more for grabbing than tearing. As Hemingway wrote accurately, makos have convergently evolved a similar body shape and coloration to the huge pelagic predator fish, such as tuna and swordfish. Mako sharks, more specifically the shortfin mako (*Isurus oxyrinchus*), live and hunt in all oceans, can grow longer than thirteen feet (4 m), and can reach speeds of over forty miles per hour (64 kmh). (The second mako species, the longfin, *I. paucus*, is lesser known.) Biologists today know that mako sharks, like great whites and a few

other related species, have not only evolved larger brains than most other sharks, but they also are warm-blooded, in part to help their muscles sustain high speeds.

The first Western scientist to identify and name mako sharks *Isurus oxyrinchus* (which means "equal tail," referring to the symmetrical lobes of the tail, and "sharp snout," because of its head shape) was a naturalist with a compelling name himself, Constantine Samuel Rafinesque Schmaltz. In 1810 Rafinesque was researching on the island of Sicily, where he wrote, "[The mako] is a fast and daring fish; its meat is not contemptible." This was not the most enthusiastic endorsement for an evening meal, but makos are known around the world as a good-tasting shark for humans.

As Hemingway accurately described in *The Old Man and the Sea*, fishers out of Cuba and the Florida Keys in the 1940s and '50s were regularly catching all species of sharks to sell for food, oil, skin, and for their fins as export to Asia for shark fin soup.

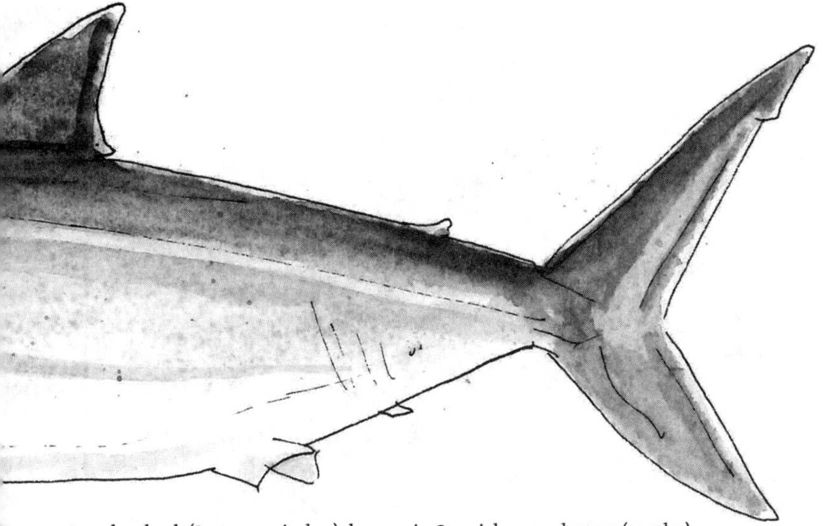

A mako shark (*Isurus oxyrinchus*), known in Spanish as *un dentuso* (toothy).

In *The Old Man and the Sea*, as the mako shark is biting off huge chunks of the marlin, Santiago, though he knows it to be futile, fights back. It is the largest mako he's ever seen. He respects mako sharks because they're not regularly scavengers. The mako hunts the largest fish, as he does. The old man manages to kill the shark with a harpoon, but then he questions both his killing of the marlin and now this mako.

"He is beautiful and noble and knows no fear of anything," Santiago thinks.

After killing the mako, the old man continues on toward the coast until inevitably another species of shark appears.

"*Galanos*," he says aloud.

For Cuban fishers at the time, *galanos* were a general type of shark rather than a given species. In Santiago's description the galanos have shovel-shaped heads, brown skin, and tails with especially larger upper lobes. The deep-water location, behavior,

Santiago stabs at the galanos sharks while trying to protect the carcass of a blue marlin (*Makaira nigricans*).

and appearance of this galano fits best for the oceanic whitetip shark (*Carcharhinus longimanus*).

Santiago respects almost all the creatures of the ocean, including the mako shark, but he hates these galanos—and not just for eating his catch. The makos approach the marlin directly with great speed. But the galanos swim underneath and attack the catch from the bottom. Thus the old man sees the galanos as cowardly. "They were hateful sharks," Santiago thinks. "Bad smelling, scavengers as well as killers, and when they were hungry they would bite an oar or the rudder of a boat."

Hemingway explains that, unlike the makos, these galanos, with their yellow "cat-like eyes," would bite the flippers off sea turtles when they were asleep on the surface, and the galanos, he

Un galano, an oceanic whitetip shark (*Carcharhinus longimanus*).

wrote, would attack humans, even if there was no blood or fish oils in the water.

In the closing lines of *The Old Man and the Sea*, Santiago limps into the harbor with only the shark-ravaged skeleton of the marlin. The meat has been stripped, torn apart by the makos and galanos, seeming to render useless the old man's heroic efforts.

After the old man stumbles up to his shack to recover, tourists at a restaurant ask about the big skeleton in the surf.

"Tiburon," the waiter says, "Esharke."

The tourist misunderstands. She and her companion think the skeleton was that of a shark, not the cause of the carcass.

In *The Old Man and the Sea*, Hemingway took us out to sea on the Straits of Florida with Santiago so we could understand more than the tourist.

Juan Fernández Crawfish

Even if you haven't read *Robinson Crusoe* (1719) or seen a movie version, you're probably familiar with the story of this legendary castaway alone on a distant island. Daniel Defoe based his fiction on the true story of a well-traveled, drunken Scottish pirate named Alexander Selkirk.

In 1704 Selkirk was on a privateering voyage on behalf of England. He felt the expedition was failing to raid enough Spanish ships. Selkirk complained bitterly around the ship that the young captain was a useless jerk and that the ship was sinking due to shipworms. When they anchored within the Juan Fernández Islands, he asked to be left there. When Selkirk looked back and saw that none of his shipmates followed him, he desperately tried to change his mind—but the captain told him it was too late. So stubborn Selkirk stayed. (The ship, by the way, did in fact sink; nearly everyone drowned due to the teredo shipworms.)

Alexander Selkirk began his new life as a castaway. He survived thanks to a herd of wild goats on the mountainous main island, which is 7.4 miles (12 km) long. He now looked out alone on the broad South Pacific from this little archipelago about three hundred and seventy miles (600 km) off the coast of Chile.

After four years and four months living alone with the goats, Selkirk watched two friendly vessels pull in. English privateer Woodes Rogers, who was in charge of the ships *Duke* and *Duchess* had decided to anchor at the island to rest and collect fresh water and food.

"Immediately our Pinnace return'd from the shore," Rogers wrote, "and brought abundance of Craw-fish, with a Man cloth'd in Goat-Skins, who look'd wilder than the first Owners of them."

Rogers rescued Selkirk and hired him on as a mate, allowing the castaway to eventually return home. The story of this real-life Robinson Crusoe has often been told. But honestly, that's kind of ho-hum, right? What is most compelling to you and me is why was the boat filled with "craw-fish"? What were *they* exactly?

Selkirk was not riding with piles of small, freshwater crustaceans, what in North America we might call crayfish. These were Juan Fernández spiny lobsters (*Jasus frontalis*), also known today as "rock lobsters." These spiny lobsters around the Juan Fernández Islands have evolved to be a unique species, but there are dozens of similar spiny lobster species crawling on ocean bottoms throughout the Pacific and the rest of the world, usually living in warmer climates than their cousins, the clawed lobsters of the Gulf of Maine and Northern Europe.

Spiny lobsters look nearly the same as the big clawed lobsters. They too have five pairs of feathery swimmerets under their wide tails and five pairs of walking legs. But the first pair of spiny lobster legs did not evolve into large grasping and tearing claws. Instead spiny lobsters have enormous, thick antennae to probe and smell and taste, even to keep them in line with others during migrations. To defend themselves, spiny lobsters have lines of sharp spikes, angled forward, that cover the base of their antennae and most of their carapace, the shell on their back. Nearly all species of spiny lobsters are also more colorful than their clawed relatives. Several spiny lobster species have enough spots, stripes, and vivid colors to rival your jazziest tropical fish. For their part, Juan Fernández spiny lobsters are mostly orange-brown with pale yellow underbellies (They, like all lobsters, turn crimson red when cooked). They can grow up to nineteen inches (48 cm) long—not including their flickering antennae, which extend longer than their entire bodies.

For those years by himself, according to one historian, Selkirk boiled and grilled his lobsters, occasionally seasoning them with allspice and loving their flavor. Selkirk was not alone—at least not in his fondness for this food.

Mariners and explorers and Indigenous cultures across the South Pacific have historically praised both the flavor and abundance of spiny lobsters. The English naturalist Joseph Banks, when sailing with Captain James Cook, claimed "sea crawfish" as one of the greatest luxuries of Aotearoa New Zealand. Banks wrote about how the Māori fishers captured them on the bottom by searching with their feet. There is a famous illustration, likely drawn by Tupaia, the navigator-priest who piloted and in-

Juan Fernández spiny lobster (*Jasus frontalis*),
also known as a crawfish or rock lobster.

terpreted for Cook, that shows a Māori man giving an accurately illustrated spiny lobster (probably *Jasus edwardsii* or *J. verreauxi*) to Banks in exchange for a fold of cloth. Some 175 years later, Thor Heyerdahl wrote of the women fishers of Rapa Nui Easter Island catching spiny lobsters at night—also using their toes.

In the time of Selkirk and Defoe, British mariners often wrote specifically of the spiny lobsters of the Juan Fernández Islands. Visiting in the 1740s, some forty years after Selkirk had left, Lord George Anson claimed these lobsters to be the best in the world and that individuals regularly weighed eight or nine pounds (3.6–4 kg); a clawed Maine lobster at a restaurant today is normally between 1 and 1.5 pounds (5–7 kg). Anson reported that so many lobsters lined the ocean bottom within their reach that when the sailors were maneuvering their launches on and off the beach, they could not help but stick the animals with boat hooks. Marine biologists have since found Juan Fernández lobsters at depths of over six hundred feet.

A century later in the 1840s, Lieutenant Frederick Walpole of the HMS *Collingwood* described how much fun his crew had catching these lobsters while at anchor beside Juan Fernández, describing the entire process of catching them from the beach or from a small boat with a hoop net. "The bottom was literally lined with crawfish of a large size; some must have weighed five pounds at least," Walpole wrote. "Catching crawfish was one of the favourite amusements of the seaman. . . . we thoroughly enjoyed both the catching and the eating. We had crawfish for breakfast, crawfish for dinner, crawfish for supper, and crawfish for any accidental meal we could cram in between."

Today, the majority of the income of the roughly seven hundred residents on the island comes from catching and selling the Juan Fernández crawfish, though there is also an expanding trade in tourism serving people drawn by the story of the real-life Robinson Crusoe. In Spanish, the language spoken on what is now named Isla Róbinson Crusoe, this lobster is called *langosta*. The spiny lobster population, however, is likely a quarter of what it was when Selkirk lived in the archipelago, and the individuals captured are on average far smaller.

The good news is that the local people passed their own regulations as early as 1935, and in 2018 Chile created an enormous marine reserve around the islands, in part to sustain and rehabilitate this local lobster fishery in the Juan Fernández archipelago. Over 40 percent of Chilean waters is now a marine protected area of some kind.

We are left with one boiling mystery: how, back in 1719, did

Daniel Defoe resist writing about how Robinson Crusoe dipped into the sea to capture delicious crawfish?

Certainly those lobsters never left the actual castaway's memory. Alexander Selkirk's earliest biographer wrote about what the pirate did when he got home to Scotland: "Day after day he spent in fishing . . . catching lobsters, his favourite amusement, as they reminded him of the crawfish of Juan Fernandez."

Killer Whale

“This peacefulness was interrupted,” wrote Teddy Seymour, “by a strange encounter of an animal kind.”

Sailing alone in his boat *Love Song* in 1986, traveling from port to port on a slim budget, he was cruising across the Gulf of Aden, bound for the Suez Canal. So he could let go of the tiller to sleep, eat, read, and do boat maintenance, Seymour was using his Aries self-steering system with its separate rudder suspended into the water. One of his most important and reliable pieces of equipment, this self-steering gear was bolted firmly to the stern of the boat. It was a thick steel contraption that used the force of the wind as well as the water going past. The steering gear had a wind vane that stuck up in the air, made of thin wood, and a rudder constructed of thick fiberglass in an airfoil shape, which stuck down about 2.5 feet (.7 m) under the surface.

Seymour had passed through a region of light winds and was making good time, sailing safely and calmly. All was going well. Earlier that day he had seen an orca, also known as the killer whale. Unlike the challenges of identifying a Risso's dolphin or anything that might fit into the name of a grampus, killer whales are one of the easier whales to identify at sea because of their glossy black skin, distinctive white patches, and their tall, thin dorsal fin, which, for males, can stick up as high as six feet, like a giant witch's hat slicing through the seas.

After dark, the same orca, or maybe another one, approached his boat.

"A mischief maker pushed *Love Song* sideways," Seymour wrote, "and made teeth marks in the Aries rudder. By the time I leaped to the cockpit, the rudder was being molested as the boat continued to be pushed and slightly tilted."

Seymour didn't estimate the killer whale's length, but orcas can grow as long as thirty feet. His boat *Love Song* was only thirty-five feet long.

"As the hull slid off the emerging monster," Seymour explained, "a part of the mass appeared; then it quietly submerged. Operating the engine and striking the engine with a hammer were two techniques used for the next half-hour to deter foul play."

This was quick thinking by Teddy Seymour. Killer whales, like all toothed whales, are highly in tune to their sense of hearing underwater: sound and echolocation are their primary methods for hunting and social communication. Engine noises are known to deter killer whales and reduce their ability to talk to each other and to find food. In the 1980s, the branch of marine biology that studied whale echolocation and their sensitivity to

Killer whale (*Orcinus orca*) approaches the self-steering rudder on a small boat.

underwater sound had only begun to emerge. Western scientists had just been identifying local dialects in several species of whales, as well as public and private clicks, pulses, and whistles underwater.

The fact that Seymour encountered an orca near the Middle East might come as a surprise, but killer whales, known as *alhut alqatil* in Arabic, are the ocean's most widely distributed animal species. Not only do orcas swim across all oceans, they also penetrate the icy waters of the Arctic and the Antarctic. Though they are rarely sighted swimming and hunting in the Arabian Sea and even into the Red Sea, they are occasional visitors to this region.

Wherever they live, killer whales are highly adapted to their particular environments. They live in localized packs to hunt regional prey, which, depending on their part of the ocean and the traditions of their given group of orcas, might be fish, penguins, seals, or even a vulnerable large whale, such as a young sperm whale or a sick humpback.

Teddy Seymour concluded his account with some humor from the safety of his writing desk back home: "Explanations submitted to account for this obnoxious behavior suggest both innocence and violence. Was the nudging an act of passion? Could it have been love at first sight? Were the teeth marks a display of affection to be viewed as a memoir? Perhaps the light-gray colored Aries rudder flashing side-to-side in the phosphorescent matter of the sea lured the animal into regarding the device as an evening snack. For the same reason that a patent log rotor is painted black—namely, shark avoidance—I accepted this incident as a suggestion to paint the Aries rudder a darker color."

Seymour's killer whale probably had no romantic attraction toward *Love Song*, despite the vessel's name, but perhaps the pale rudder moving through the bioluminescence might indeed have looked like a fish or a flipper for a hungry killer whale. The rudders of these devices are quite hardy, but I suspect if that orca—with each tooth longer than your finger—had really wanted to take a full, confident chomp, it would've left more than teeth marks.

After that night with the orca, Seymour and *Love Song* continued on to sail through the Mediterranean, across the Atlantic, and drop anchor in his home harbor at St. Croix in the Virgin Islands. Seymour, who grew up in Yonkers, New York, became the first African American person to complete a solo circumnavigation of the world.

Teddy Seymour was not, however, the first or the last sailor to come hull to head with a killer whale. Recently orcas off the coast of Portugal and Spain have been making the news. In 2021, according to *SAIL* magazine, more than fifty boats reported close encounters initiated by killer whales, half of which resulted in damage that required heading into port for repairs.

Louisiana Shrimp

Martha Field, writing under the pen name of Catharine Cole, was the first woman to be a full-time reporter for the New Orleans newspaper the *Daily Picayune*. In the 1880s her stories, especially her travel essays, earned her the status of a celebrity throughout Louisiana, and her byline became known nationally. Often adventuring by herself, Field sloshed through alligator-infested bayous, rowed in precarious canoes, traveled on horse-drawn buggies, and observed the world from sailboats, steamboats, and trans-Atlantic liners. Soft-spoken and perceptive, she reported from all over the state of Louisiana and filed accounts from as far away as Chicago, Washington D.C., and from several European cities.

In July of 1888, Field published a story about her expedition touring the barrier islands of the Gulf of Mexico, particularly in Barataria Bay off the mouth of the Mississippi River. She chartered a small schooner, named the *Julia*, which was painted green and run by a couple and their young daughter. As a family, they hunted birds to sell for decorations on women's hats in France. Although Field's writing sadly followed in line with the racism of her region and her newspaper, she was an outspoken advocate for women's equality—for a (white) woman's right to work, to travel, to own property, and to vote. Her stories serve today as a window into human life, marine animals, and the rapidly shifting coastlines of southern Louisiana.

Aboard the *Julia* that summer, they fished, captured seabirds, and explored ashore. Toward the end of the voyage, Field and the family anchored off Timbalier Island West. Today this island and Timbalier East are a quarter of the size they were then due to sea level rise and hurricanes, but even then they were thin, barely inhabited islands with just a few trees.

Field observed, "Shrimp are to be had by the barrelful just for seining, and on Timbalier and other islands equally flat and dreary the Chinese make their camps over the old haunts of [the pirate Jean] Lafitte. . . . It is a great industry, that, of drying shrimp and salting down sea trout, and the Chinese understand both to perfection. The shrimp swim in shoals directly on the surface of the water, and are caught in cast-nets."

On another of her trips, Field sailed to one of the larger shrimp-drying camps near Grand Isle, which was home to at least a half-dozen large communities of fishers. Here she met with one of the women who lived in that community and learned that the Chinese and Filipino people lived and worked together in "rush-thatched" homes on wood platforms that they had built onto pilings driven by hand into the marshes and low island areas. The women of the community wove the cone-shaped seine nets, and the men tossed the nets from wooden boats with lugsails dyed red. They carried their shrimp home in baskets. Then the entire community boiled the shrimp and spread them on the platforms in the sun to dry.

Field wrote: "The method of shrimp-drying is apparently simple enough. It is said, however, the Chinese have a secret formula. Perhaps it could be resolved down into simple doses of saltpeter. When dry, they are salmon-red, and as hard as bullets. They are then raked up and put into white canvas bags. The shells are threshed off, either by walking on the bags or by beating them on a board."

Several pockets of people from the Philippines and then from China had settled in this region in the 1800s, some of whom had planned for jobs on the plantations after the Civil War and Emancipation, but then rejected this work and the treatment. Some who migrated to the Louisiana Coast had first settled the San Francisco Bay region, where they had found shrimp stocks had already been overfished, and the government policies were restrictive and racist. Perhaps the most famous of these nineteenth-century immigrant communities in Louisiana was composed mostly of Filipino Americans and known as Manila Village and Saint Malo. Neither of these villages exist today because of land loss and the hurricanes that relentlessly devastate this part of the world.

Shrimp are decapod invertebrates, meaning they have ten legs and no bone-based skeleton. They're closely related to lobsters; the major difference is that shrimp are smaller and evolved to more regularly swim up in the water column. Shrimp have gills under their carapaces, and five pairs of swimmerets under

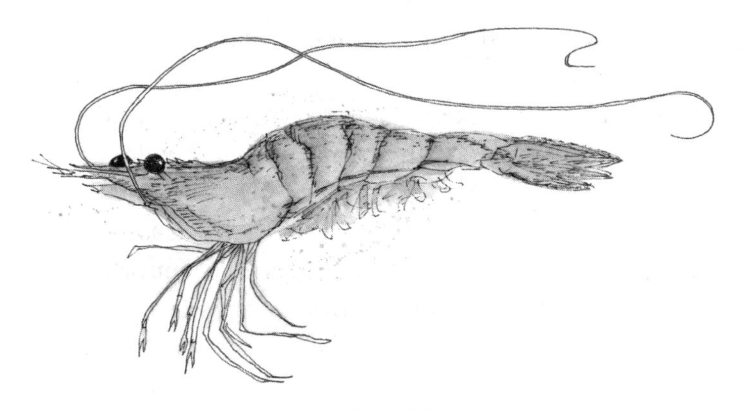

A Louisiana, or white, shrimp (*Litopenaeus setiferus*).

their tails, which they use for swimming, paddling backward. The tail is the part that humans eat, because that is where most of the animal's muscle is. Shrimp have eyes on stalks, but, as with spiny lobsters, their primary method for sensing their world is the pair of antennae, which are like long thin whips that can grow over twice as long as their bodies.

Marine biologists have identified more than two thousand separate species of shrimp worldwide, living in practically all oceans, at all depths, temperatures, and salinities. When Martha Field visited this part of the Louisiana Coast, the fishers probably captured mostly white shrimp (*Litopenaeus setiferus*), also known today as common shrimp or gray shrimp, since they are a light bluish-gray color in life. White shrimp can grow to be eight inches (20 cm) long, the females a bit longer than males. White shrimp favor shallow, warm, lower-salinity areas. They swim around coasts, bayous, and islands of Louisiana all year but are most common in the warmer summer months. Today, along with the similar brown shrimp (*Farfantepenaeus aztecus*) that prefer deeper waters, white shrimp account for the majority of the shrimp caught off the coast in the Gulf of Mexico.

With the growth of refrigeration technologies, the development of coastal trawlers with engines, and the building of highways, fresh and frozen Louisiana shrimp became a more popular product than dried shrimp. Today, because it is labor-intensive and the demand is smaller, shrimp are dried by only a few fishers, such as those of the Louisiana Dried Shrimp Company, based on Grand Isle. The father of the owner of this company learned the craft from his Chinese neighbors in an area of the island that was then known as "China Town." Fishers no longer dry them in the sun on long platforms, nor do they do what was known as "dancing" on the shrimp to remove the shells. Now they dry them inside with electric heaters. You can still buy little packets of dried US shrimp in local stores in Louisiana and around the country, and the product is still sold to China and other overseas markets. One of the longest-running dried shrimp businesses was sold by Bob Hoy, a Chinese American PhD physicist who left his research-faculty position to take over the family dried shrimp business in New Orleans. His father had started the international Quong Sun Company in the 1920s, later renamed Gulf Food Products. Hoy said that he doesn't eat shrimp that much now because it's too expensive. But when he was younger, "we used to just eat them dry, like peanuts."

For her own taste, Martha Field preferred fresh-cooked fish over dried shrimp, but the site and smell of shrimp always reminded her of the Gulf and the barrier islands. In one of her dispatches from Houma, Louisiana, the sight of piles of the "great pink crescents" got her thinking of the large kettles managed by the Filipino fishers who boiled the "pale grey trophies" until they turned a crimson red.

Piles of shrimp sent Field into revelry of her time fishing and sailing in the Gulf. "Somehow, everything else in pretty Houma paled beside the pink fires of those sun-dried shrimps. I wonder what it is in me that irresistibly draws me seaward! Why, the very sound of a conch is enough to set me thinking on a voyage; the smell of shrimp makes me long for a sail around Little Caillou. It must be that I had a sea king for an ancestor."

Mother Carey's Chicken

Minnie Lawrence, five years old and the daughter of the captain, was sitting in the cabin with her mother looking out the window. Their whaleship, the *Addison*, was just to the northwest of the Galápagos Islands. The crew had recently caught a sperm whale, and the men were processing the blubber. It was a messy, loud, and dangerous time for a kid to be up on deck, but Minnie didn't seem to mind missing out, especially because there was so much to see out this window at the stern. The butchering of the two small whales had attracted prolific marine life all around their ship.

Minnie's mother, Mary Chipman Lawrence, kept a detailed journal over the course of their voyage from New Bedford, Massachusetts to the Pacific. It's one of only a few full journals that remain from the hundreds of wives of whalers who went to sea in the 1800s. Mary was one of three wives of three different whaling captain brothers from the same Lawrence family. So Minnie was one of a handful of cousins who spent years at sea as kids on whaleships.

On this day, March 15th, 1857, Minnie's mother ended her journal entry: "We saw at one time six sharks following the ship, with the pilot fish ahead of them and no end to the skipjack and the albacore, while on the surface of the water and flying around are hundreds of Mother Carey's chickens. The air is almost black with them. Minnie says, 'How many pretty things we see on the ocean, don't we?'"

"Mother Carey's chickens" was a common nickname for storm-petrels, the smallest of open ocean seabirds. Individuals of the smallest of the small, the least storm-petrel (*Hydrobates microsoma*), can weigh as little as half an ounce (14 grams). That's the weight of a child's handful of paper clips, barely more than a chickadee.

About twenty-two species of storm-petrels (family Oceanitidae and Hydrobatidae) fly above all the world's oceans. Most of them are dark, ashy gray with a range of white markings and a short, black, tube-nosed beak. The storm-petrels that Minnie and her mother saw that day in such a large flock following the ship could have been a group of over a half-dozen species that transit through that part of the Pacific, including the Wilson's storm-petrel (*Oceanites oceanicus*), which is one of the most numerous and most widely distributed seabird species on the planet. Maybe Minnie and her mother were witnessing a flock of wedge-rumped storm-petrels (*Oceanodrama tethys*), a spe-

Wedge-rumped storm-petrel (*Hydrobates tethys*), known to sailors as a "Mother Carey's Chicken."

cies that nests in cliffs and volcanic crevices on the Galápagos Islands.

Gulls, albatrosses, prions (known as "whale birds"), and other large species that feed on hunks of fish and other ocean carrion are usually the birds that follow whaleships. Storm-petrels usually eat the smallest of ocean organisms found right at the surface, tiny crustaceans like krill or amphipods, but also larval fish and squids. Yet Minnie and Mary's observation was not unprecedented. Storm-petrels follow ships of all kinds and will gather around a dead whale. Perhaps they feed on bits of blubber and, when following ships, maybe pick up small prey stirred up in a vessel's wake.

Minnie's observation and the way her mother described the event is not only notable for its small bit of environmental history—the abundance of two types of tuna and a large flock of storm-petrels at this time and in this location—but also because of how pleased she and her mother seem to have been to see these storm-petrels and to equate them (along with the sharks, pilot fish, and tuna) with "pretty things."

Storm-petrels have had a mixed reputation in the folklore of sailors and coastal English-speaking communities in northern Europe and the United States. The word *petrel* probably comes from the story of St. Peter, who walked on water in the Bible, because this kind of seabird, particularly storm-petrels, appear to tiptoe over the waves with their thin, dainty webbed feet while they are feeding, seemingly skittering over the surface as if they're treading air. The nickname "Mother Carey's Chicken" might come from the Latin *Mater Cara*, meaning Virgin Mary, the protector of sailors. Or maybe it's from a really old fairy tale about a farmwoman named Mother Carey, who tended the entire ocean like it was her field. Storm-petrels can also emit a little chittery screech that you can hear from astern, which is vaguely chicken-like.

Whatever the derivation, the nickname Mother Carey's Chicken was widely used in Minnie Lawrence's time. In the 1830s, for example, the painter-naturalist John James Audubon, who seemed to have loved storm-petrels more than any other seabird, liked to feed them off the stern (when he wasn't trying to shoot them for specimens to draw). Audubon created paintings of two species of storm-petrels in the North Atlantic, and he wrote of them as "Mother Carey's Chickens" in his sea journal. He later published lush descriptions of storm-petrels in flight, "full of life and joy," and he, along with at least one of his captains, saw storm-petrels as brave survivors, harbingers warning sailors of storms and rough weather.

Many mariners and fishers, however, saw it the other way around, that storm-petrels actually brought the bad weather, as if they are tiny witches. Herman Melville, who had been a whaler himself in these same waters a little over a decade before the Lawrence's voyage around the Galápagos, wrote a short story called the "The Encantadas, or The Enchanted Isles" (1854) that includes a description of storm-petrels in this area with an association of dread: "The stormy petrel or Mother Cary's chicken sounds his continual challenge and alarm. That this mysterious hummingbird of ocean, which had it but brilliancy of hue might from its evanescent liveliness be almost called its butterfly, yet whose chirrup underneath the stern is ominous to mariners as to the peasant the death-tick sounding from behind the chimney jamb."

The idea of storm-petrels as storm-conjuring witches persisted well past the mid-1800s and into the modern day. Poet and folklorist Charles Leland, who collected sailor songs and crafted new ones, wrote down the following chantey in 1895 (which I think Minnie would've loved to have sung while stomping around the quarterdeck as the sailors hauled on a halyard):

With the wind old Mother Carey,
Yo ho oh!
Churns the sea to make her dairy:
Yo ho oh!

When you see a storm a-brewin,'
Yo ho oh!
That is Mother Carey's doin':
Yo ho oh!

When you see Mother Carey's chickens,
Yo ho oh!
Then look out to catch the dickens!
Yo ho oh!

A half-century later, Sir Francis Chichester, an English sailor who set a record for the fastest solo circumnavigation in 1967, often saw these storm-petrels in the Southern Ocean, home of the most furious waves and winds on Earth. Chichester wrote: "That evening a big swell began running in suddenly from the west; big, I would say 50 feet. There were a number of Mother Carey's Chickens about, which nearly always forecasts a storm, whatever meteorologists may say to the contrary. I could see them picking things out of the water while on the wing, but not what the things were."

Today, storm-petrels in some coastal communities are often less favored. In Nova Scotia, for example, some citizens think they are smelly, magotty birds, and treat them differently than other small seabirds, such as puffins. This is likely a remnant of the witchy folklore about storm-petrels. Before electricity, historical residents of some of the coastal island villages in the North Atlantic used to insert wicks into storm-petrel carcasses, using their bodies as lamps, since the birds have layers of fats and oils to keep them warm and nourished out at sea.

As for Minnie sailing with her parents aboard the *Addison* in the Pacific in the late 1850s, she celebrated four birthdays out at sea. She visited with Inuit people in the Arctic, lived for weeks in Hawai'i, and toured around the Gulf of California. She sailed across the Southern Ocean and twice around Cape Horn. Minnie saw more "pretty things" out at sea before she turned nine years old than most people do in a lifetime.

New Zealand Sea Lion

St. Clair Beach, on the Te Waipounamu/South Island of Aotearoa New Zealand, is a long stretch of fine white sand. It's desolate compared to a beach in New Jersey or on the Italian Riviera, but by Kiwi standards it's a very busy destination with decent surf, and beyond the swimmers a parade of penguins, sharks, and dolphins passing through. For the people at St. Clair Beach, there's a short line of coffee shops and restaurants as near to the sand as safety allows. And at the edge of the paved walkway on the way down to the ocean is a strikingly realistic statue of a light gray sea lion and her tiny brown pup snuggled to her side. The statue, built with fiberglass materials and techniques similar to the making of surf boards, is titled "Mum: A Legendary Sea Lion." The statue was commissioned after this sea lion's death, which locals believe occurred sometime around 2010.

To explain why a group of community members would erect a statue like this, we need to go back about a thousand years or so.

Around the year 1200 CE, as Polynesian voyagers continued their open ocean crossings and centuries-long settlement of the South Pacific, the mariners made their final major stop on what would be the last large inhabitable land mass on Earth that had yet to be occupied by humans. Going back and forth from their home islands, they settled here in a series of *iwi*, or tribes, communities who would become what we now call collectively the Māori people of Aotearoa New Zealand. The Māori soon discovered that these large islands did not have any native land mammals other than two species of bat. To feed themselves, they had brought rodents and plants with them, and in their new land they captured seabirds, such as penguins, shearwaters, and albatrosses, and hunted large, flightless land birds, especially enormous moa, which had evolved and thrived in this mammal-free archipelago.

The Māori settlers also turned to *kaimoana*, or seafood, such as fish, eels, lobster, and abalones (*pāua*), as well as to the marine mammals that spent part of the year on the beaches of Aotearoa, animals that had never encountered humans before. There were several species of pinnipeds here to hunt—seals and sea lions—including one species that is found only in these waters, the *whakahao* or *rāpoka*. In a later time, English-speakers called them Hooker's sea lions (*Phocarctos hookeri*) after the English botanist Joseph Hooker. Today, to English-speakers, they're mostly known as the New Zealand sea lion. Like all pinnipeds, they are sexually dimorphic: the large brown males can grow to

Female, *kake*, and pup of a New Zealand sea lion (*Phocarctos hookeri*), or *whakahao*.

be nearly 11.5 feet (3.5 m) long and weigh almost 1,000 pounds (450 kg). The females, with cream-colored fur coat, grow no more than 6.5 feet (2 m) long and weigh about 350 pounds (160 kg). The Māori language further specifies the males as *whakahao* and the females as *kake*. New Zealand sea lions, like all pinnipeds, need safe beaches to haul out to mate and raise their young. Females, doing the raising on their own, sometimes take their single pup nearly a mile (~1.5 km) inland and into high brush and forest to leave the young ones while they go back out to sea to fish.

In recent years, female New Zealand sea lions, who forage deeper than males, have been tracked diving almost two thousand feet (600 m) down into the subantarctic waters to capture their preferred prey of deep-water fish, squids, octopuses, and crabs. New Zealand sea lions, especially the males, will also, if they can catch them, eat penguins or cormorants. Sea lions use their large eyes and extraordinarily sensitive whiskers, known as vibrissae, to hunt in the dark and to try to escape their one known predator, the great white shark.

A favorite prey of New Zealand sea lions are New Zealand arrow squids (*Nototodarus sloanii*), or *wheketere* in te reo Māori.

Due to centuries of Māori subsistence hunting, archaeologists and biologists believe the New Zealand sea lion was eradicated from the main three islands of Aotearoa (North, South, and Stewart) by about 1700. The species survived only by way of breeding colonies on remote subantarctic islands, such as on the Auckland Islands or at Campbell Island.

In the 1800s, the remaining populations of the New Zealand sea lions were pushed still further to the brink of extinction by American, British, and French settlers and commercial hunters who arrived with steel harpoons and then firearms, killing pinnipeds on the most remote beaches. Americans and Europeans killed sea lions for the skins and oil, and occasionally for food. This hunting was conducted on a large scale and proved devastating to all pinniped populations in this part of the world. In addition to the local right and sperm whale populations, the sea lions and fur seals were one of the main reasons that Europeans began to frequent and then settle in Aotearoa New Zealand. In time, there were so few pinnipeds in these waters that it wasn't worth the trouble for colonial hunters to seek them out.

Similar to changes in perceptions of whales, there has been a dramatic shift in the relationship to sea lions within different cultures of humans, particularly those of industrial nations, in a relatively short period of time, partly because cheaper lubricants, fabrics, and other resources have been found. In other words, to put it crudely, once some cultures didn't rely on the given animals for food and products, they opened up more toward conservation.

It took less than a century for many communities around the world to switch from hating and fearing sea lions to loving and wanting to protect them. Consider that in 1845, Captain Washington Peck, who was sailing near the Chincha Islands off the coast of Peru (and advocating for the rights of Chinese laborers as he inhaled the toxic scent of cormorant guano), still harbored deep distaste for the sight of a pinniped. Captain Peck believed the fur seals and sea lions off the coast of South America to be "horrible" and "monsters" and "the compelled agents of some diabolical spell or inevitable doom."

In 1874 French sailor François Raynal published a well-read account of his time shipwrecked on the Auckland Islands—the remote islands far to the south of Aotearoa New Zealand. Raynal explained how important marine mammals were to their survival, as well as adding some useful biological observations of the New Zealand sea lion. Raynal was quite impressed with the animals, the "amphibia," but still saw their ferocity, especially of the males.

"Pressing one against the other," Raynal wrote, "their eyes glowing, their nostrils expanded and snuffing the air with a loud noise, their lips, trembling with rage, turned upwards, these monsters opened wide their enormous jaws, surmounted by long, stiff mustaches, and displaying the most formidable tusks. Every moment they flung themselves upon one another, and bit and gnawed, tearing away great shreds of flesh, or inflicting gashes whence the blood flowed in abundant streams."

Raynal described how one sea lion took a bite out of the bow of their boat. The accompanying illustration of the male in his book has a face, comically in retrospect, that is nearly identical to a terrestrial lion.

Yet not long after, in 1893, hunting seals and sea lions was banned in the waters of Aotearoa New Zealand. This was primarily for economic reasons, but the ban revealed a collective understanding of their vulnerability. Throughout the 1900s, New Zealanders along the shorelines only witnessed scattered males occasionally hauling out on southern beaches, such as St. Clair, during a few winter months a year. Females and their young ceased coming anywhere near the main islands. In 1978, pinniped preservation was extended more forcefully and widely by the country's Marine Mammal Protections Act, by which sea lions became legally protected throughout Aotearoa New Zealand.

Then in 1993, exactly a century after the first legislation, something miraculous happened. For the first time in probably more than a hundred and fifty years, a female New Zealand sea lion hauled herself out on a beach of mainland Aotearoa, at the mouth of the Taieri River, a few miles below St. Clair beach. She gave birth to a pup. A lot had changed since the hunters abandoned the area. Now in the 1990s people were excited to watch and protect these "sea bears."

And "Mum," as she became known, became a pioneering sea lion. She swam her pup to a still more remote beach just east of St. Clair. Over the following years, Mum returned several times and went on to give birth to eleven pups. The New Zealand author and naturalist Neville Peat wrote a beautiful book called *Coasting: The Sea Lion and the Lark* (2001), which imagines Mum's annual migration, the birth of her pups on the mainland—a "wildlife phenomenon" of "sea lions coming home"—and traces the marine ecology of this species. Mum's female pups returned to the South Island beaches to mother their own. Likely other female New Zealand sea lions followed them, too.

Today, between ten and twenty pups are born in this region every year, with a possible resident population of nearly three hundred sea lions. Now there are over a dozen different pupping sites on the lower coast of the Te Waipounamu/South Island and Rakiura/Stewart Island.

The New Zealand sea lion, however, remains Earth's rarest species of sea lion. They remain classified as endangered by the IUCN, and their numbers continue to decrease, with a 2021 estimate of about twelve thousand individuals. New Zealand sea lions remain threatened by development and human disturbance in areas where they haul out to mate and raise their young. Overfishing has reduced their available prey, and the sea lions—especially the females and young, which prefer the same regions as the fishers—often get killed or injured in the nets of fishing vessels trawling for squids, despite the establishment of marine protected areas, and required improvements in fishing gear designed to reduce these incidents.

On St. Clair beach, beside the statue of Mum and her pup, is an interpretative sign written by the New Zealand Sea Lion Trust, founded in 2003. The sign extols the legacy of Mum's reintroduction of the New Zealand sea lion species to mainland shores. The sign teaches visitors about giving the animals a proper distance. At the bottom is a note from Edward Ellison, a leader of the local

Ngāi Tau *iwi*, the tribe who holds this sea lion in high regard and is actively involved in its current protection:

TE HOE KAKARI O TE WERA, HE WHAKAHAO
"The belligerent enemy of Te Wera was a sea lion,
Te Wera and his seventy men warriors were routed by the
angry sea lion." So it is fitting that such an exalted species
is seen to be regenerating along our coast. It had status
and values that were admired by our forebears.

Noddy

Despite the name-calling and disparagement of their intelligence, including by the likes of Charles Darwin, noddies are aesthetically magnificent, and they have been immensely valuable, even lifesaving, for human fishers and navigators.

Looking a lot like a dark tern, but more closely related to a gull, the brown noddy (*Anous stolidus*) is the largest and the most widespread of the three or four global noddy species. The brown noddy transits across all oceans, preferring tropical and subtropical seas and islands for roosting and nesting. Their only breeding island in the United States is on one spot in the most distant of the Florida Keys, but their worldwide populations seem reasonably stable today, with nesting populations on islands, for example, in Hawai'i, the Phoenix Islands, the Red Sea, and on selected islands of Indonesia. Brown noddies also nest, among other locations, off both western and eastern Australia, in the Galápagos Islands, and on islets of the Seychelles Islands in the northwestern Indian Ocean. Noddies have beautiful black eyes with thin white crescents around the edges set within plumage that gradually fades from a slate or ashy brown body to a white cap of pale feathers. They eat squids and fish, usually by gliding low and nabbing flying fish from the air or pecking at the surface to capture prey, but they occasionally patter on the sea like storm-petrels or belly-flop onto the water to reach down below. Noddies have been observed hitching short rides on the back of sea turtles and pelicans, and they often go out fishing on moonlit nights.

Brown noddy (*Anous stolidus*).

Charles Darwin failed to see the beauty of these seabirds on his outbound passage aboard the HMS *Beagle* in 1832. Usually seasick when out of sight of land, he was primarily interested in getting off the ship to study the rocks ashore. Early in their world circumnavigation, Captain Fitzroy anchored beside St. Peter and St. Paul, a little archipelago of low-lying islands in the equatorial Atlantic. Although he would not find any guanay cormorants in this part of the world, Darwin wrote about the white guano, how it had been identified to be so valuable off the west coast of South America. "We only observed two kinds of birds," Darwin wrote, "the booby and the noddy."

"The former is a species of gannet, and the latter a tern," Darwin continued. "Both are of a tame and stupid disposition, and are so unaccustomed to visitors, that I could have killed any number of them with my geological hammer. The booby [the brown booby, *Sula leucogaster*], lays her eggs on the bare rock; but the tern [likely two species of noddy, the brown and the black, *Anous minutus*] makes a very simple nest with seaweed.

By the side of many of these nests a small flying-fish was placed; which, I suppose, had been brought by the male bird for its partner. It was amusing to watch how quickly a large and active crab (*Graspus*), which inhabits the crevices of the rock, stole the fish from the side of the nest, as soon as we had disturbed the birds."

Young Darwin was hardly the first to ascribe a lack of intelligence to noddies and other seabirds. James F. Stephens had a few years earlier published a similar assessment of noddies in Shaw's *General Zoology* (1825), which Darwin seems to have internalized: "They are said to be a very stupid race of birds, and to allow themselves to be knocked on the head without attempting to remove from the place; they are usually of very dark and somber colours." The scientific genus name for noddies, *Anous*, means silly or clueless in Greek, and the brown noddy's species name, *stolida*, means slow or dull in Latin. Stephens in 1825 wrote that the common name "noddy" comes from "their apparent stupidity flying into ships" as well as allowing themselves to be caught—although, he reported, they put up a scratching fight once captured.

At sea around this time, in 1826, John James Audubon was sailing aboard a ship bound to London from New Orleans through the Straits of Florida. In his journal, Audubon drew a gorgeous, detailed, and life-sized illustration of a brown noddy. One of the crew had captured the animal when it was sitting on one of the ship's booms.

"I know nothing of this Bird more than what Our Sailors say," Audubon wrote, "that it is a Nody and frequently alights about Vessels in this Latitude."

In England during Darwin and Audubon's time, *noddy* was a slang word for "sleepy," but also an adjective for simple and foolish, even idiotic, which went back to at least the 1500s. And for nearly as long, this particular bird was known as a noddy in the English language, as revealed in early European exploration narratives.

A more recent suggestion of the derivation of the name noddy, which is more charitable if likely incorrect, is that the name refers to the ways that noddies bow as part of their mating. They do have a rapid head-up, head-down behavior display—both the male and female move their beaks up and down very close to each other—as if they're continually, forcefully agreeing with the other: very much nodding.

What the lubberly nineteenth-century naturalists and the early European merchant sailors were missing about noddies, as well as other seabirds, is that not only were they slow to avoid human hands because they had never learned to be afraid of people, but that ocean-reliant Indigenous communities throughout the world had for millennia relied on seabirds, especially noddies, to help them find fish on the surface, as well as to help them navigate when out of sight of land.

In the Indian Ocean, for example, noddies regularly join feeding frenzies over schooling fish, capturing the smaller fish, squids, or crustacea scared up by tuna and dolphinfish. They don't carry their fish home in their beaks, like terns, but swallow their food and then regurgitate it back to their young ashore. This is likely what Darwin saw on St. Peter and St. Paul—the regurgitants filled with flying fish that the crabs quickly nabbed (like good seabird and fisheries researchers). For centuries before modern fish-aggregating technologies and sonars, fishers off the Maldives, for example, used noddies specifically to locate schools of skipjack tuna.

In the Pacific, noddies were some of the most useful birds for the early navigators, especially those sailing among low island atolls, because not only do flocks of noddies go home to roost each night, they tend to range farther from shore than day-foragers. One master navigator from the Gilbert Islands, Teeta Tatua of Kuria, told sailor-anthropologist David Lewis in the late 1960s, “Birds are very useful up to twice the sight range of an island from a canoe. The sight range of land is about ten miles [16 km] and that of the birds twenty [32 km]. The birds which are most significant are terns and noddies.”

The more solitary frigatebirds often go quite far out to sea, and tropicbirds, who are usually seen alone or in pairs offshore, have even less of a predictable flight path in terms of land, but at sunrise or near dusk, a flock of noddies flying inbound or outbound draws a neat path in the sky toward land. Modern studies have confirmed the range reported by Teeta Tatua, although brown noddies in the eastern Indian Ocean have also been observed flying some 125 miles (200 km) from their colonies to fish.

So take that Darwin, you noddy.

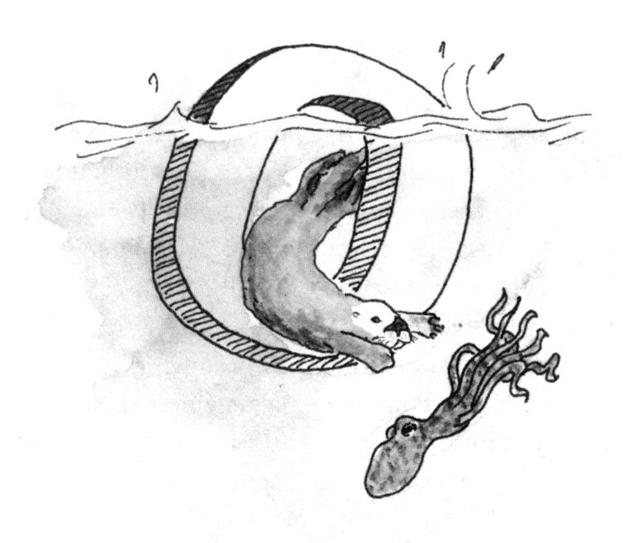

Octopus

Marilyn Nelson wrote a short, powerful poem in 2019 titled "Octopus Empire." The second and final verse is this:

> Now scientists have found
> a group of octopuses
> who seem to have a sense
> of community, who
> live in dwellings made of
> gathered pebbles and shells,
> who cooperate, who
> defend an apparent
> border. Perhaps they'll have

a plan for the planet
in a millennium
or two. After we're gone.

Nelson was inspired to write this poem after reading about research conducted from Jervis Bay, just to the south of Sydney, Australia, which identified cooperative practices in a group of gloomy octopuses (*Octopus tetricus*). Marine biologists had thought these gloomy octopuses, like all octopuses, were solitary for their whole lives. But it appears some individuals have formed into a few small communities, occupying little heaps of broken shells and dens that the animals engineered. Nelson used this discovery to imagine an octopus community with a collective intelligence and wisdom that might care for the Earth better than humans have so far. The poet evoked these cephalopods to teach people about our moral failure as the supposedly most intelligent species on the planet, presumably better armed with a larger brain.

You've come a long way, octopus—at least in the eyes of people. In Western literature for the last two thousand years, the octopus has been portrayed as the devil, the kraken, a creature that is horrible, dumb, and evil. Pliny the Elder, the Roman author of the bestiary *Historia Naturalis* (c. 75 AD), explained that the octopus is in most respects "stupid," although clever with how it moves shells around to form a lair and attract fish. Pliny relayed, however, that "no animal is more savage in killing a man in the water: it struggles with him, embracing him with its tentacles, swallows at him with its many suckers and pulls him apart; it attacks shipwrecked men or men who are diving." This general impression continued up through the fiction by the French authors Victor Hugo and Jules Verne in the late 1800s. Hugo's octopus in *Toilers of the Sea* (1866) is "the monster," an enigmatic embodiment of evil and everything of which humans are most terrified. "If terror be an object," Hugo wrote, then "the octopus is a masterpiece." Reverend Harvey's giant squid fit right within this writhing, dangerous depiction.

This popular perception of the monstrous octopus, often confused in a tentacular inky cloud with krakens, sea snakes, serpents, and giant squids, slowly shifted in Western literature and marine biology in the mid-1900s, notably influenced by a now little-known American biologist, inventor, and boatbuilder from the Chesapeake Bay by the name of Gilbert Klingel.

Klingel's relationship with octopuses began when he somehow convinced the American Museum of Natural History in New York City and the Natural History Society of Maryland to fund the building of a wooden replica of the *Spray*, the boat in which Joshua Slocum had been first to sail alone around the world. Klingel and a colleague planned to sail around the Caribbean to study reefs and land reptiles. But in their new boat, named *Basilisk* (after the lizard that can skitter over the surface of the water), they did not get far. The two young men shipwrecked in rough weather on the reef off Great Inagua at the southern end of the Bahamas. Undaunted, they set up camp ashore and recorded all they could, both on the island and underwater around its reefs. Several years later, Klingel went back to Great Inagua for further study, and in 1940 he published *The Ocean Island: Inagua*, in which he wrote his chapter "In Defense of Octopuses."

The inspiration for his appeal began one day when he was walking on the seafloor off Great Inagua using a diving helmet. Klingel was coming up an underwater ravine and was about to put his hand on what he thought was a yellow rock, when, by chance, he noticed eye slits. He watched the "rock" ooze slowly away, like hot wax, its skin changing color from pebbly yellow to red, then to white. Though the octopus's head was about as big as

a football, it slithered out of sight down into a crack in the reef that was smaller than a teaspoon.

Klingel spent many days underwater observing the octopuses of the reef, including the large one that he first mistook for a rock. There are some three hundred known species of octopus that live in a range of marine habitats all over the world, from Antarctic waters to equatorial coasts, from intertidal zones to deep sea, and these species include a few different ones that live in the reefs around the Bahamas. Klingel's octopus that day might have been a large individual of the Caribbean reef octopus (*Octopus briareus*). He estimated its total arm span as five feet (1.5 m) from tip to tip.

In his "Defense of the Octopuses," Klingel wrote of the animals' uncanny abilities for camouflage, in which they not only change their color, but their shape, and even their skin texture.

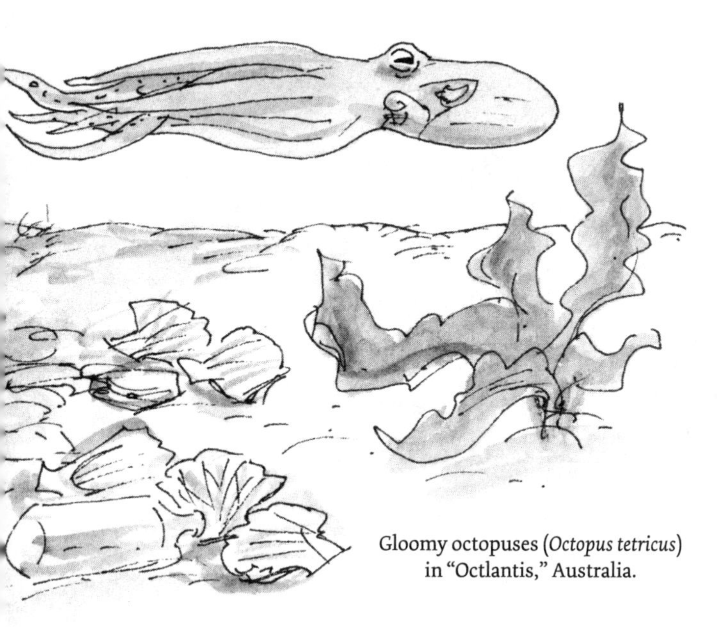

Gloomy octopuses (*Octopus tetricus*) in "Octlantis," Australia.

Klingel watched the octopuses' clever strategies to capture crabs, and he observed how they stored uneaten shellfish just outside their dens within the reef.

At one point when Klingel was observing one large octopus, he decided to see what would happen if he gave it a little poke with a stick, to see how the skin color might change in response. All at once, the octopus grabbed the stick with its arms and let it go, sending the stick floating to the surface, while at the same time squirting ink into the water before jet-propelling away. Klingel smelled a "fishy musk" that seeped into his dive helmet. He was surprised to see the color of the ink was not black, but more a dark purple that faded into a "somber shade of azure."

Despite the inking, Klingel wrote that these animals have "been the unknowing and unwitting victims of a large and very unfair amount of propaganda, and have long suffered under the

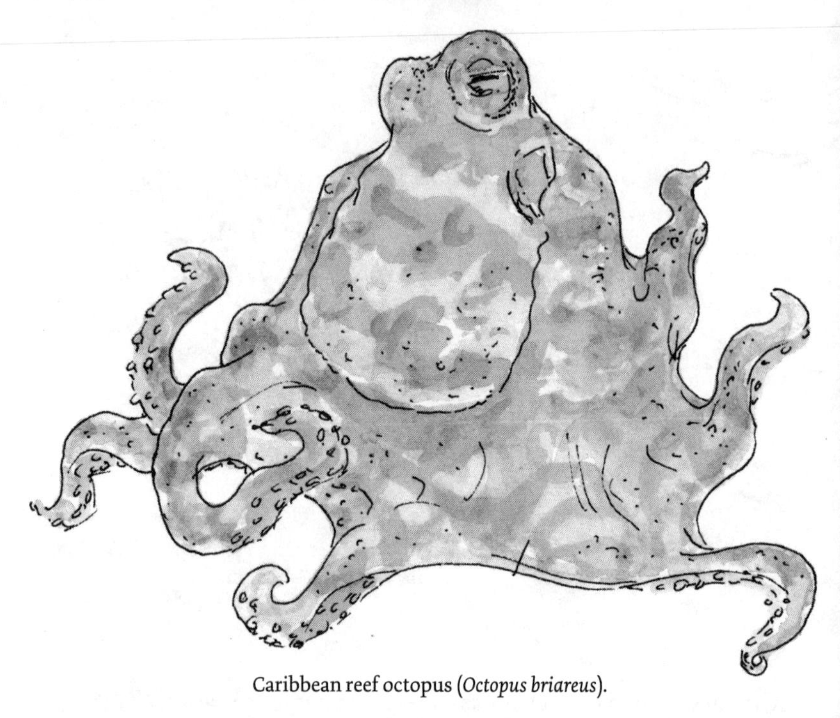

Caribbean reef octopus (*Octopus briareus*).

stigma of being considered horrible and exceedingly repulsive." Klingel declared instead that octopuses were "among the most wonderful of all earth's creatures."

In this way, Gilbert Klingel was ahead of his time in expressing his deep respect for these animals, written several decades before the best-selling book *Soul of an Octopus* (2015) by Sy Montgomery, the award-winning documentary *My Octopus Teacher* (2019, produced by Craig Foster), and Nelson's poem "Octopus Kingdom" (2019). A half-century earlier, Klingel had marveled at the intelligence of octopuses, which are "only" invertebrates, yet these cephalopods use tools, appear to play, exhibit intense curiosity, and have the ability to learn in ways equivalent to mammals, maybe even at the level of some of the primates. Klingel wrote in 1940 something that foreshadows Nelson's "Octopus Empire": "There is reason to believe that they are the most keen-witted creatures in the ocean and had they developed an opposable thumb and fingers instead of suckers with which to manipulate various objects the entire course of the earth's existence might have been altered."

Octopus intelligence has been proven time and time again. At the very least, by human measurements of smarts, octopuses are the most intelligent of the invertebrates. This is all the more extraordinary because octopuses mate only once during their short life span, about one to five years in the wild, and they die after mating. This means the adults never see their babies emerge from their eggs, so there is no parent-to-child transfer of knowledge or practice. Veterinary researcher Greg Lewbart quipped in 2015: "I had a professor who said if octopuses ever evolved into terrestrial creatures, they'd rule the world."

So today Marilyn Nelson can comfortably write a poem using gloomy octopuses, without irony, to teach human morality, to inspire human behavior, as she writes about that region off Australia where marine biologists observed communal octopus behavior, an underwater community they dubbed "Octlantis."

"Reading the article gave me some hope for the planet," said Nelson, helping her get out from under "the gloom cloud of what seems to be the planet's bleak future."

Otter

With her pony-tail under a baseball hat, captain-owner Wendy Kitchell points at a sea otter swimming just off the port bow. Her tour boat, powered by an eco-friendly electric engine that is entirely silent when in neutral, has not yet even left the dock. The cameras come out and people smile. Floating on his back, the otter looks over with his big triangular nose and long whiskers, then rolls around and dives underneath the surface.

"Too cute!" someone gasps.

"Everyone see that?" Captain Kitchell says (the tour offers a money-back guarantee if passengers do *not* see an otter.)

As on nearly every trip, though, the passengers will go on to see dozens of otters this morning—playing, courting, fighting, swimming, diving, feeding, dozing, and teaching their young. The captain explains that by law and by practice, they give the sea otters as much space as possible, never approaching them directly, but the animals have become quite accustomed to their quiet boat. They are in Elkhorn Slough, the central estuary of Monterey Bay, California, where about a hundred and fifty sea otters live year-round, which local ecologists believe might be the carrying capacity for this area. It is likely the highest concentration of the southern subspecies of sea otters anywhere.

Captain Kitchell steers the boat off the dock. The passengers watch sea lions, cormorants, pelicans, and herons as they quietly glide up into the slough and under the highway bridge. Most of the passengers, though, are really looking for more sea otters.

When they do see a few more otters, floating on their backs and grooming themselves with small, hand-like paws, the naturalist on the boat, Natalie Rossi, explains that, unique among all other marine mammals, sea otters do not have a significant blubber layer to keep them warm in this cold part of the Pacific. Instead, they have two layers of fur. Their underlayer of over 650,000 hairs per square inch is the most dense fur of any animal on Earth. A typical human, for contrast, has about 100,000 hairs—over our entire head. Sea otter fur is tantalizingly soft and silky to human skin and exceptionally cozy and warm.

Natalie goes on to say that sea otters, especially up in Alaska, can grow nearly five feet (1.5 m) long. The males are larger than females and can weigh a hundred pounds (45 kg). Sea otters are coastal animals, often found floating around in estuaries and in beds of giant kelp, within which they can dive down to about a hundred and thirty feet (40 m), collecting urchins, abalones, clams, snails, and crabs. In addition to their dense fur, sea otters stay warm by having an exceptionally high metabolism; they eat more than a quarter of their weight every day.

As the boat passes two sea otters floating on their backs off the starboard side, Captain Kitchell tells the group to watch for them using a second clam to smash open a hard shell. In the outer bay, they might use a rock. Sea otters have pocket-like folds of fur under their front legs in which they may store their personal hammer.

This boat ride took place in autumn of 2021. Less than a century before, there would have been no chance of seeing a single sea otter anywhere in the slough or anywhere in Monterey Bay, since the animals were considered extinct throughout California. Because of the colonial fur trade, biologists and environmental historians believe that by the early twentieth century, less than 5 percent of the pre-contact sea otter population remained.

Sea otter (*Enhydra lutris*) with urchin.

As with the New Zealand sea lion, to explain how this animal population got here, it's helpful to go back several centuries. Although scientists recognize about thirteen species of otters around the world, the other twelve prefer freshwater and sleep on land every night. The one true saltwater sea otter, *Enhydra lutris*, has historically lived around the coastal rim of the North Pacific, from northern Japan all the way around to Baja California. (Biologists today recognize three subspecies of sea otter, separated geographically.) Western-style archaeology, early colonial accounts, and the oral traditions of First Nations confirm that humans have been hunting sea otters for thousands of years on both sides of the North Pacific, including the Indigenous people of the Ainu, Kamchadal, Aleut, Sugpiaq, Tlingit, Alutiiq, Haida, Nuu-chah-nulth, Makah, and all the way south to tribes in California, such as the diverse groups of Esselen peoples around Monterey Bay and the Chumash in southern California. Middens from the Calendaruc people, who lived around Elkhorn Slough for some eight thousand years, reveal that they hunted sea otters among other marine mammals.

The colonial fur trade began in the mid-1700s. Russian explorers and hunters began capturing seals and sea otters around the Aleutian Islands and then over to the coast of Alaska. Sea otters fetched an exceptionally high price for Russian and British mariners, especially when first trading in China, where people also valued sea otter fur for coats and other garments. Chinese merchants had themselves, since the 1400s, been trading for sea otter furs from the Ainu people to the north. The Russian hunters killed—and traded at great profit—far more sea otters than the Indigenous people had ever done in previous centuries. The Russians also forcibly, sometimes cruelly, employed Indigenous hunters to catch the otters for them, including Aleutian men who were renowned for their skill with spears and boats.

As colonial hunts and trading continued in the 1800s in the North Pacific, involving Spanish, French, and American hunters, sea otters became more and more scarce. Russian settlers in California tried some early conservation practices, but few people adhered to them with the price still so high. Colonial mariners sailed their ships farther and farther south, nearly to Mexico, to find sea otters. Interest in otter fur expanded in Europe, too. Few of the colonial traders cared about the long-term protection of sea otter populations, or, for that matter, the Native American people who lived there. By the mid-1800s, the animals were nearly impossible to find.

The early 1900s was the nadir for sea otter populations, at the brink of global extinction. Fortunately, demand decreased, and the sea otters began to recover naturally. In 1911 the major countries involved agreed to reduce the hunting of some North Pacific marine mammals, such as fur seals and sea otters, which was the first international marine conservation agreement of any kind.

In the 1930s, word got out that a small population of surviving sea otters had been living along the coast to the south of Monterey Bay, reigniting interest in these animals in California. By the 1960s, American managers began a program to capture sea otters in the Aleutian Islands, put them in ships and airplanes, and reintroduce them to their former ranges to the east and south. Meanwhile, sea otter conservation rode the wave of public interest in protecting ocean life, due in part to the work of Eugenie Clark, Archie Carr, and Rachel Carson. The vulnerability of cuddly sea otters was specifically featured in Scott O'Dell's young adult novel *Island of the Blue Dolphins* (1960). The appeal of sea otters has increased, as the animals live reasonably well in public aquaria, so tourists can see these puppy-like animals up close. Federal environmental laws in the 1970s, such as the Endangered Species Act, the Marine Mammal Protection Act, and local laws reducing gill net fishing in their coastal habitat further assisted sea otter recovery.

All this, quite reasonably, is not on the mind of the tourists on the electric boat today, although Natalie shares with them that the global sea otter population is growing, and many of the coastal sites the southern sea otter occupied before European contact, over 10 percent, have been reclaimed by the animals. Marine ecologists have found that in the Monterey Bay sea otters are a keystone species, crucial predators that regulate the health and diversity of the coast and estuaries.

The story gets complicated, however, in terms of social justice, especially farther to the north. Indigenous communities along the Pacific Coast had their own conservation traditions and tribal regulations well before the arrival of colonists. The Tlingit, who call the otter *yáxwch'*, are one of the Native peoples along the shores of what are now Alaska and British Columbia. Tlingit people have hunted the sea otter for millennia and crafted the pelts into ceremonial robes, fashion elements, and bedding, especially for chiefs and others of high rank. This continues today, cultivating and continuing their hunter's knowledge and Native environmentalism. So the imposition of American and Canadian laws by the governments of the people who devastated the coastal ecosystems in the first place are pretty hard to swallow—especially since tribal councils were never consulted in any of the early conservation management.

Nor were Native communities ever included in the introduction programs. The recovery of sea otters is welcomed by Indigenous communities on some levels, such as the return to historical relationships with the animal and the benefit of its value as a resource. They are glad as well as for tourism income earned by Native-owned wildlife-watching boats. But on other levels, the sea otter's return is a double-edged sword. Sea otters are major predators of shellfish in kelp beds, so their return in many locations might be a large part of the reason for crashes in the abalone, clam, and urchin fisheries that had provided food and income that many Native communities had come to depend upon for several decades.

Treaty Manager Peter Hansen, of the Kyuqot/Checleset First Nation, said publicly in 2003, "In Checleset where sea otters were first transplanted in the 1960s, we have noticed declines in most shellfish. We have not been able to harvest urchins, geoduck, clams, scallops, or abalones for the last ten years, and the children have not had the opportunity to have these foods."

In this region today, sea otter populations have recovered so well that 90 percent of all the world's sea otters now live in coastal Alaska. Tourism is a large industry, but so is fishing and harvesting food. First Nations do have a say once again in conservation, and they are the only American people allowed to hunt sea otters, with a major say in setting their own quotas.

Back at the dock in Elkhorn Slough, California, the happy, tired ecotourists step off the quiet boat and onto the pier, wave goodbye to Natalie and Captain Kitchell, eager to scroll through their photos of sea otters. This local population of sea otters is thriving, and the tourists and the residents—fishers, biologists, and business-owners—reverentially give them their distance.

Paper Nautilus

We'll start this story with William Wood, a naturalist from England. He was a member of the Royal Society with a particular expertise on shellfish. It wasn't all his fault.

In 1807 Wood published *Zoography; or, The Beauties of Nature Displayed*. This three-volume bestiary of sorts, exquisitely illustrated, included his personal favorites for the most fascinating animals, plants, and minerals on Earth. It offered an especially glowing account of the paper nautilus, also known as the argonaut or the paper sailor.

Wood explained the theory that the earliest humans had learned the very idea of sailing from these animals. He wrote that the eight-armed "sailor within" sits inside a papery white shell "marked with elegant ribs that run towards the keel." Wood said that the animal spreads out a membrane at the end of two specialized arms to form small sails, gliding his shell across the water. At other times, presumably in calms, the paper nautilus spread out his arms to row. In the illustration in *Zoography* to accompany this story, the two "sails" are presented held up high on two arms. The other arms, thin and wormy, seem to reach down out of the shell to refresh themselves daintily in the water. In the background is a coastal scene with traditionally rigged human sailboats.

Wood went on to explain that whenever approached by a human, the paper nautiluses would quickly dive, making them difficult to capture and nearly impossible to observe at sea. The last part is true, but almost everything else is fantasy.

William Wood's false understanding of the physiology of this animal, to be fair, went back to at least ancient Greece in 300 BCE. Aristotle's descriptions of these creatures sailing across the surface had subsequently led the famed Swiss zoologist Carl Linnaeus in the early 1700s to name the animals with the genus name Argonauta after the Greek myths of the hero-sailors, the Argonauts who had voyaged aboard the ship *Argo*.

Two decades after William Wood, the French naturalist Jeannette Villepreux-Power began to examine paper nautiluses more carefully and scientifically. Born in France, Villepreux-Power had gained renown as an embroiderer in Paris. In the one known photograph of her, she wears a gorgeous dark embroidered shawl over a hoop dress and, at the back of her dark hair, an embroidered net. She presents her profile to the photographer with a thin-lipped expression that seems to say, "When will this photo shoot be over so I can get back to my lab?"

When Villepreux-Power moved to the coast of Sicily, she began to satisfy her long-held interest in marine biology, geology,

Female paper nautilus (*Argonauta* sp.); the male (*left*) grows to only about 1" long (2.5 cm).

and marine illustration. Noticing paper nautiluses in the bay and receiving specimens from fishers, she began to wonder how she could carefully observe these animals. William Wood had written, after all, that they were impossible to watch on the ocean.

Although their outer shells can be similar in size and shape, paper nautiluses should not be confused with their distant cephalopod relative—the thick-shelled, multi-tentacled chambered nautilus (subclass Nautiloidea). Chambered nautiluses are fixed inside their shells, like abalones or mussels, but, as Villepreux-Power and William Wood knew, the paper nautiluses (*Argonauta* spp.) are actually free-moving octopuses living inside a paper-thin, nearly transparent shell. The shell of the largest of the four species of argonauts can grow to sixteen inches (41 cm) across. That's longer than a bicycle helmet.

Beginning in 1832 and continuing for a full decade—before much of her collection was destroyed in a shipwreck—Villepreux-Power studied paper nautiluses. She started by trying to solve the debate as to if these animals make their own shells. Scientists at the time wondered if the argonauts instead found their shells, like hermit crabs, or maybe even parasitically killed the original maker of the shells. Villepreux-Power also wanted to find out how these animals reproduced, and if argonauts did indeed sail or row across the surface.

In order to study paper nautiluses alive in her harbor off the Mediterranean Sea, Villepreux-Power designed wood cages that were anchored underwater. For her small coastal laboratory, she designed hoses to pump seawater into wood and glass enclosures. Many historians therefore credit Villepreux-Power with the invention of the modern aquarium, then called a "vivarium." And since she recommended some of these strategies for raising fish, she also advanced European aquaculture.

Her experimentation proved that the octopuses do indeed make their own shells from a young age, and they use their sail-like appendages to build this shell as they grow.

Villepreux-Power also found that the rosy-purply octopuses inside the shell are only females. The shell serves in part as an egg case. The "dwarf males," as they are now called, look entirely different and are tiny, as much as six hundred times smaller than the females. Biologists today know that a minute male has no shell, and when he visits he only leaves behind a single arm with his sperm attached. Villepreux-Power never figured out this part, nor did any of the Western scientists of the early 1800s. Most believed at the time that this little male arm was some kind of a parasitic worm that happened to live inside the paper nautilus's shell.

As to if paper nautiluses catch the wind, Villepreux-Power wrote that she never once saw a paper nautilus raise her two webbed arms in the air. Villepreux-Power proved through observation that the tissue membranes on the two arms, the "sails," serve primarily to build and repair the shell, as well as to cover it while the animal is swimming.

In a profession antagonistic to women at the time, Villepreux-Power was not afraid to defend her research. When a French naturalist named Sander Rang stole her ideas and tried to take credit for her discoveries, Villepreux-Power published a rebuttal in 1856, which included in part, "Not only are his observations wrong, but his drawings are incorrect, and the way they are presented perfectly demonstrates that Mr. Rang did not study with that thoroughness and perseverance that are necessary to arrive at an exact knowledge of the manners and habits of this animal."

Biologists today believe that paper nautiluses are the only octopuses that fabricate and live within shells and among the few octopuses that live in the open ocean rather than along the sea floor, like the Caribbean or gloomy octopus. Modern biologists confirm that the argonaut's membranes do not, unfortunately for the legacy of William Wood or for writers such as Jules Verne, propel the animal by the wind—not even a little bit. In fact, paper nautiluses move themselves under the water by jetting, swimming quickly and purposefully by expelling water out of their funnel, in the same way as do other octopuses and squids.

The last part of our story about the evolving Western understanding of the paper nautilus takes place in the Pacific. In 2006 an Australian marine biologist and curator named, appropriately, Julian Finn, captured three argonauts off the coast of Japan. As a guest in a local lab beside Okidomari Harbor, he and a colleague observed the animals in a 317-gallon (1,200 liter) tank. Finn then released them back into the harbor for sessions where he could scuba dive beside them and observe them in open water.

Dr. Finn discovered that paper nautiluses have evolved a way to rock their shells back and forth to trap air bubbles from the surface before they dive down. The animals then regulate their buoyancy with this air pocket as they descend—similar to how Finn adjusted his air within his buoyancy vest as he was watching the animals.

"I've studied argonauts for many years," Finn once said in his office in Australia, "and I've looked at thousands of shells in museums, and I've gone through old texts and I've read up on the old writings, but it wasn't until I actually got an argonaut in the water that I truly saw the true marvel of these animals."

Parrot

The most famous photograph of Captain Michael Healy, the first Black captain in the US Coast Guard, prominently features the man with a pet parrot perched on his rocking chair as they sit on the quarterdeck of the cutter *Bear*, which is docked in blustery Alaska. The parrot appears somehow not to be squawking, "Polly want a scarf."

This photograph was taken around 1895. Today we consider a captain's parrot a cliché, a stock character: a parroty of itself. This is the story of how two famous writers played with the parrot character in their sea novels. And did these fictional birds reflect actual life at sea, such as that of Captain Healy and his parrot?

Charles Johnson featured a ship's parrot in his novel *Middle Passage* (1990), which is a Gothic sea adventure set in the 1830s that follows a recently free Black man, Rutherford Calhoun, who finds himself aboard a ship of horrors. Unaware of what he is getting himself into, but eager to escape circumstances in New Orleans, Calhoun steals the ship's papers of a cook who is passed out cold in a bar. Calhoun is nearly stymied, however, by the man's parrot, who first warns "Bad move," and then loudly squawks "Thief! Thief!"

With the stolen papers, Calhoun sneaks on the ship, although the cook, named Squibb, and his parrot make it back before the vessel sets sail. Once out on the Atlantic Ocean, Calhoun and Squibb—and the parrot—become friends. The bird, it turns out, is as hard a drinker as the cook. The parrot likes to joke with the sailors about their sex lives, and at night the bird sleeps on Squibb's belly, like a gull on a floating whale. Toward the end of the story, the parrot sheds tears along with Squibb.

In *Middle Passage* Rutherford Calhoun is a survivor who spins the yarn of a disaster at sea. He regularly references other classic sea stories, like that of a novelist whose work is crucial in any august analysis of ship's parrots: Robert Louis Stevenson.

Stevenson's *Treasure Island* (1883) gave readers what is now the most renowned of parrot characters. Long John Silver, also a ship's cook, shows young Hawkins his pet parrot, a female, which he gives sugar snacks and calls Cap'n Flint. The bird not only swears and barks sailing orders but repeatedly shrieks "Pieces of eight!," a key to the story. Silver explains, "Now that bird . . . is, may be, two hundred years old, Hawkins—they lives for ever mostly; and if anybody's seen more wickedness, it must be the devil himself. She's sailed with England, the great Cap'n England, the pirate. She's been at Madagascar, and at Malabar, and Surinam, and Providence, and Portobello . . . and to look at her you would think she was a babby [*sic*]."

It turns out that Stevenson and then Johnson did not imagine parrots on ships out of nowhere. Large, brightly colored, vocal birds live almost exclusively in tropical climates, so when the early European explorers sailed to warmer latitudes, they collected parrots as souvenirs and specimens for friends, collectors, and scientists back home.

Historians believe Christopher Columbus, for example, brought back two Cuban Amazon parrots for Queen Isabella. In 1676 explorer and pirate William Dampier observed red and yellow "tame Parrots" along the Caribbean coast of Mexico. "They would prate very prettily," Dampier wrote, "and there was scarce a Man but what sent aboard one or two of them."

One cruel strategy sailors used for collecting these parrots involved seizing one and pinching the animal until it cried out. Since parrots are socially protective birds, others would soon fly in, tricked into being caught themselves. Once the sailors had the parrots aboard, they built cages to keep the birds safe for the passage home. The birds did not, however, always survive their journeys at sea. On the Atlantic Ocean in 1826, for example, the log-keeper of the brig *Cicero* reported dryly, "At 9 AM took F[ore] T[opgallant] Sail. Lost a parrot overboard belonging to Mr Delgado."

For centuries Western mariners collected parrots and other tropical birds. When the sailors returned to Europe or North America, to major cities such as Paris, London, and Boston, merchants ran stalls that specialized in selling the birds they collected. Sailors and their merchants often sold the parrots through newspaper advertisements or at taverns. First lady Mary Washington had a pet parrot, as did President James Madison a few decades later, both birds surely delivered by ship.

The word *parrot* can be a general term. Biologists still debate the groupings of these birds, which represent hundreds of species. As a popular name, especially in historical sailors' accounts, parrots included macaws, cockatoos, parakeets, lovebirds, and a variety of other large, strong-beaked birds, most of whom eat seeds, fruits, and nuts as the majority of their diet. In part to help with rain forest camouflage, parrot species are often mostly green, like the parrot Cap'n Flint in *Treasure Island*. Captain Healy's parrot, by the photo, was perhaps a yellow-naped parrot (*Amazona auropalliata*). In *Middle Passage*, Charles Johnson clarifies that the cook's bird is an African gray parrot (*Psittacus erithacus*). These parrots are nearly all gray, but with a thick black beak, white skin around their yellow eyes, and brilliant red tail feathers. Johnson chose his parrot species well, not just because the ship was bound to Africa, but because African grays

African gray parrot (*Psittacus erithacus*) on a sailor's shoulder.

were common pets at sea, and this species of bird has long been known for its exceptional intelligence, capacity to learn, and voice-mimicking ability. African gray parrots can have direct conversations with humans, as shown by cognitive research beginning in the 1970s by Professor Irene Pepperberg with "Alex," an African gray parrot who over three decades built a vocabulary of over one hundred vocal labels and who proved to have smarts comparable to a human toddler.

Collecting African gray parrots, especially along the Gabon River in the West African country now called the Gabonese Republic, was apparently a thing for mariners in the 1800s. In 1859 a sailor aboard the *USS Marion* wrote that from the coast of Africa his ship's company had gathered some fifty pet parrots, likely African grays. Some were kept in cages made from netting and some even in old tea kettles punched with holes. The sailor on the *Marion* wrote, "James Cummings, does not growl quite all the time, now, as he did before he bought a parrot in Gaboon. His whole attention & time is taken up with his noisy bird, which he appears to love as a parent would his child." Two decades later, a sailor on the *USS Lancaster* wrote that the crew "secured some magnificent specimens of the African gray parrot," also from the Gabon River. Naval officer Henry Augustus Wise (a.k.a. "Harry Gringo") wrote a short book in 1860 titled *The Story of the Gray African Parrot*, with a wonderfully long subtitle that tells you pretty much all you need to know—although I've added a few notes in brackets so you can keep it straight. The subtitle reads, *Who* [the parrot, known as Polly or Baby] *Was Rescued by the Little Sailor Boy in the River Gaboon; How He Whistled, and How He Talked, including His* [the Parrot's] *Great Battle with the Monkeys, which lasted six weeks: and How He Behaved During the Awful Shipwreck* [saving the life of his owner], *Together with Some Account of His Latter Days*. The parrot hero retires to live a long, peaceful, chatty life outside of Boston with the sailor's mother. We can't put Henry Wise on the same shelf of novelists with Charles Johnson and Robert Louis Stevenson, but it is quite a story nonetheless and further fixed the parrot as a comic character in sea stories.

Today, parrots are as iconic a symbol of sailors and pirates as a striped shirt or a peg leg, and numerous parrot species are endangered or vulnerable. Healy's yellow-naped parrot is now critically endangered. African grays are classified as endangered due to the international pet trade (now traveling by plane) and the increasing loss of their forest habitats. Parrots do not live as long as the fictional Cap'n Flint from *Treasure Island*, but some individuals can live to over forty years in captivity. In the novels *Middle Passage* and *Treasure Island*, the parrot's ability to talk, to mimic human speech, makes the animal useful as a watchbird—just as the parrot's intelligence and chatty nature were surely part of its appeal as pets for actual mariners at sea, including Captain Mike Healy, for at least four hundred years.

Thus a sassy, chatty parrot on a sailor's shoulder, a classic supporting actor in sea fiction, has its perch on historical fact.

Pilot Fish

A shark about six feet (1.8) long appeared alongside the merchant ship *Whitaker* while the passengers and crew were eating dinner on Monday, March 20, 1738. They were in the heart of the Atlantic, nearly two weeks out from Gibraltar, westbound for Savannah, Georgia. Among those looking over the side to watch the shark was the twenty-three-year-old George Whitefield, a preacher who would go on to be one of the most famous and dynamic Methodist orators of his time. He was on this voyage as a missionary. Benjamin Franklin's press went on to publish Whitefield's journal from this trip across the Atlantic, in which the young man described the sight of this shark and, more significantly, how five little fish swam around the shark's mouth. Whitefield knew these as pilot fish—and he ascribed great meaning to their behavior.

The voyage so far had been eventful. Whitefield was seasick for multiple days. He had watched an electric storm and suffered through bouts of heavy weather and adverse winds. Once better himself, he'd been visiting those ill with fever and teaching the soldiers on board how to read and write. He'd been throwing playing cards and "bad books" overboard, and he gave a sermon one day on deck against swearing that, he reported, brought the crew to tears. Whitefield officiated one marriage aboard, and had gone on the two other ships with which they sailed in company to perform two other marriages (although for one he had to close his bible in the middle because the groom was giggling too much). Whitefield and the passengers and crew had seen dolphins leaping around their ship, and later in the trans-Atlantic voyage they would see grampuses and dolphinfish. But it was only this vision of the shark and the pilot fish that truly moved the preacher in his journal to see a Divine lesson for people in marine life.

Whitefield reported that the pilot fish looked like large mackerels as they swam in company with the shark. "What is most surprising," he wrote, "though the Shark is so ravenous a Creature, yet let it be never so hungry, it never touches one of them. Nor are they less faithful to him: For if at any Time the Shark is hooked, these little Creatures will not forsake him, but cleave close to his Fins, and are often taken up with him."

Whitefield saw this as enough proof of a Christian God to refute an atheist. He was moved in his journal to compose a moral to the relationship between the pilot fish and the shark: "*Go to the Pilot-fish, thou that forsakest a Friend in Adversity, consider his Ways, and be abashed.*" In other words, those of you who have ditched your friend (or God), however sharkish, when things get tough, just look at the pilot fish for inspiration of good moral behavior: stay faithful and be a friend to the end.

George Whitefield would not be the only person at sea to wonder about the behavior of the pilot fish and the meaning, the metaphor, of this tiny animal living peacefully with what would seem to be its predator.

Pilot fish (*Naucrates ductor*) are not to be confused with remora (family Echeneidae), the fish that suction themselves, literally attach their heads, to sharks, whales, and sea turtles. Pilot fish are independently swimming bluish-silvery fish with thick black vertical stripes. One species lives in all oceans, including in the Mediterranean Sea, preferring tropical and temperate waters. Pilot fish typically align themselves along the leading edge of the head or dorsal and pectoral fins of dozens of species of sharks, including oceanic whitetips, mako, and the mammoth whale sharks. Pilot fish also swim alongside large rays, sea turtles, and other larger, slower moving fish. Pilot fish will also collect under large mats of sargassum, driftwood, and slow-moving ships and sailboats, like Thomas Albro and Minnie Lawrence's whaleships and Teddy Seymour's *Love Song*. Young pilot fish favor swimming within jellyfish tentacles and drifting seaweed.

In Whitefield's 1700s, and for well over a century to follow, naturalists, sailors, and passengers remained confounded about the relationship between pilot fish and sharks. Some believed that the pilot fish took a leadership role, like human pilots for ships coming into a harbor, which was the likely derivation for the name (*ductor* is Latin for "guide" or "leader"), meaning that they believed the fish actually guided sharks toward their prey.

In the 1830s Harvard-educated Richard Henry Dana Jr., who spent two years as a merchant seaman, wrote that a pilot fish swimming alongside the hull meant that a shark was on the way. Others reported the pilot fish's devotion, like Whitefield, while others reasoned this to be simply an opportunistic relationship—a few pilot fish benefiting from the bully and enjoying the protection and the scraps—the pilot fish being just as happy to enjoy the shade of a ship as protection, a bigger host.

Sailing around the same time as Dana and Darwin, the naturalist Frederick Bennett sailed as a ship's doctor on a British whaleship in the South Pacific. Bennett described the pilot fish in detail. He wrote of the pilot fish's eye color (iris is hazel, with an outer ring of "gold-yellow"); their teeth (single row around upper and lower jaw); average size (between six and fourteen inches [15–36 cm], although modern biologists have recorded individuals that are two feet [.6 m] long); and the number of dark stripes (about five, not including those on the head and tail; we now know these stripes fade as they grow older, eventually dis-

Pilot fish (*Naucrates ductor*).

appearing entirely). Bennett confirmed that these pilot fish will accompany ships, are commonly found around sharks, and eat a variety of smaller fish and invertebrates.

Dr. Bennett carefully considered the nature of this relationship with sharks, wondering if the larger benefited from their presence. He observed that the pilot fish could catch and eat food by themselves, so they did not *need* the sharks. He also very much doubted earlier stories that sharks needed the pilot fish to guide them. Then again, Bennett explained, he had personally witnessed something astounding in 1832 on a passage from India: a lethargic shark passed by bait that the sailors had placed on a hook. The shark did not go after the bait until one of his two pilot fish zipped out first to examine it and swam back to the shark, who turned around and then snatched up the bait—to his own demise. After the crew killed the shark, the two pilot fish "expressed the greatest concern, almost leaping out of the water," and remained with the vessel for weeks, even attending a small boat of theirs that went to and from another ship at sea. Of one of these pilot fish, Bennett wrote, "So devotedly did it attend upon what it might have believed to be its lost shark, as to lead the officers of the *Thomas Grenville* to remark that we had a Pilot-fish *painted* on the rudder of the boat."

A careful reader of Bennett was author and whaler Herman Melville, who subsequently took up the question of the relationship between pilot fish and shark in a little poem he wrote later in his life titled "The Maldive Shark" (1888), in which Melville saw, like Reverend Whitefield, moral lessons in the relationship—but with a far different conclusion. Melville wrote:

> From his [the shark's] saw-pit of mouth, from his charnel
> of maw
> They [the pilot fish] have nothing of harm to dread,
> But liquidly glide on his ghastly flank
> Or before his Gorgonian head;
> Or lurk in the port of serrated teeth
> In white triple tiers of glittering gates,
> And there find a haven when peril's abroad,
> An asylum in jaws of the Fates!
> They are friends; and friendly they guide him to prey,
> Yet never partake of the treat—
> Eyes and brains to the dotard lethargic and dull,
> Pale ravener of horrible meat.

For Melville, the pilot fish was hardly a symbol for the good Christian. This animal, though a friend and a guide, was more a wily, opportunistic survivor: a self-serving ally and beneficiary of the slovenly shark's protection. Nor would the Reverend Whitefield have appreciated the way that Melville wrote about pilot fish in two of his earlier novels, *Mardi* and *Moby-Dick*, in which the author's pilot fish are also opportunistic, quick to leave a dying or doomed host to find a better one—the heretical opposite of Whitefield's moral of everlasting loyalty from the pilot fish.

Twenty-first-century marine ecologists believe that sharks don't seem to get too much out of the partnership with pilot fish,

Pilot fish around a blue shark (*Prionace glauca*).

except perhaps fewer parasites on their skin. The relationship is more what ecologists refer to as symbiotic, living in close association that is mutually beneficial, although the pilot fish seems to get more out of their time together. (Pilot fish sometimes even feed from the shark's excrement.)

In a curious coda to this story about the pilot fish/shark relationship, in 2013 Portuguese marine biologists and veterinarians flew out to the Azores Archipelago. Here they dragged a tuna head in the water, thereby attracting blue sharks and their pilot fish partners. With a barbless hook and line the scientists captured eighty pilot fish, which they flew home and placed in tanks and monitored with excruciating care. By car, the team then delivered pilot fish to several different public aquaria throughout Europe and, by aircraft, to other facilities as far away as Budapest and Dubai. The pilot fish in their new homes were quarantined, kept in a safe cage for acclimation, and then released into the larger tanks to join with their potential symbionts, being several different species of sharks and large fish.

Within a week, in each and every aquarium, every single pilot fish was eaten by the resident sharks.

What do you make of that, Reverend Whitefield?

Quahog

The quahog (pronounced `kō-hog) is a clam. This animal is a bivalve invertebrate with gills, a liver, a heart, and two oval shells connected with a thick hinge. Quahogs dig themselves in just under the ocean bottom and send up two straw-like siphons: one tube to inhale sea water for oxygen and microscopic plants; and the second tube to exhale that water, waste materials, and gametes. Since quahogs prefer warm, muddy, and sandy bottoms in at least partly salt water, they are common in shallow bays and the mouths of rivers.

Quahogs are found along the North American east coast and the mainland edges around the Gulf of Mexico. Marine biologists today identify two species. The northern quahog (*Mercenaria mer-*

cenaria) ranges from the Canadian Maritimes down to Florida. The southern quahog (*M. campechiensis*) can be found from Florida south around the Gulf of Mexico, to nearly the Yucatán Peninsula. Where the ranges of the two species overlap, a large percentage of quahogs are hybrids. The northern quahog's shell is generally smooth on the outside and usually has that famous purple edging and swirls on the inside. The southern species has rough, more pronounced ridges on the outside of the shell, and its inner shell is usually entirely white.

Male clams of both species at a given time of year send expansive clouds of sperm out of their siphons into the water, which stimulates the females to send out eggs. According to one study in 1980, one large mature female quahog can send out 16.8 million eggs in one spawning season. Fertilized eggs move through a few planktonic forms, the last of which is a little crawling and swimming "peliveliger" phase which finds a suitable bottom, holds on with threads, then digs into the mud or sand with its wee foot and goes on to develop into a full-sized, mature quahog within a few years. Adult quahogs typically grow up to four inches (10 cm) wide, although a boy in the summer of 2020 along the coast of Rhode Island dug up a clam that was a full six inches (15.2 cm) across. Northern quahogs that manage to survive human fishing, pollution, and predation by fish, sea stars, and gulls can live to be at least forty-six years old.

In 1758, over two centuries after first European contact with Native peoples in the American northeast, Carl Linnaeus coined for this particular clam the first scientific name of *Venus mercenaria*. "Mercenaria" is from the Latin for wages, a hired worker, a mercenary. Linnaeus must have read about how beads made from this clam's shell, "wampum," had been used as a form of exchange, a type of money, and perhaps thought it the same situation as the cowry shell.

Northern quahog (*Mercenaria mercenaria*) with siphons sticking up out of the mud.

Unfortunately, this connection between the quahog clam, wampum, and currency has been oversimplified over the years. Wampum, specifically beads made by Native artisans from quahog and other shells, such as whelks, was used as a substitute for coins when trading with colonists. But this was only a historically late and very brief means of exchange in North American history, a fraction of the animal's meaning to Native American communities.

To start, the word *wampum* is anglicized from an Algonquian word, *wa"pa"piag*, which meant specifically a string of shell beads that were only white, but colonists reduced the word to simply mean all beads made from shell. And Europeans anglicized the word *quahog* from names for this clam like that of the Narragansett, who call the animal *poquaûhog*. For Native American tribes in this part of the world, the quahog is primarily a source of food and spiritual connection. Historically, people crafted the quahog into trowels, scrapers to carve wood boats and bowls, into utensils for eating, and even into a set of tweezers.

Beyond appreciating quahogs for food and tools, Wampanoag and other tribes have used the shells to make beads, as well as other art forms. Oral histories and archaeological midden remains reveal that Wampanoag ancestors and other Algonquin peoples have eaten quahogs and made beads from the shells for over four thousand years. The women in these coastal communities, especially during the first centuries of European colonization, became the gatherers of shellfish to feed their families and the artisans who created wampum.

Paula Peters, SonkWaban, a writer and historian of her Mashpee Wampanoag tribe in Massachusetts, explained in 2022, "My whole life I've been engaging with the quahog, digging the quahog, harvesting it, eating it—raw, in chowder, stuffed, as part of a clam bake. It is sustenance. The quahog is also something we celebrate, as sacred. And we recognize the shell that carries that life as sacred."

Peters said that opening up the shell is always a joyful surprise, to see each unique pattern of white and purple, and this shell is still used to make tools and art. Although even within Wampanoag communities today, the word *wampum* is used sometimes as a slang for money, historically this was never a direct synonym.

Wampanoag artisans still create wampum in a variety of

forms today. To make the beads, pieces of the quahog shell are carefully broken, filed, and smoothed into a tubular shape, then painstakingly drilled, often by hand with a bow tool, into each end of the bead to make the hole. Even if using both the purple and white parts of the quahog shell, only a few beads can be made from one animal. So hundreds, sometimes thousands of quahogs must be harvested to make a wampum adornment for the wrist or neck or for the beading of a long sash or belt. Native artisans weave and arrange the mixture of white and purple colors into a design with a specific meaning or story to tell, such as a creation myth or a community history. Wampum garments are created for sacred reasons, for storytelling, and some of these individual works of art have lives of their own for particular communities, similar to the life of a document like the Declaration of Independence or the passing down of a family photo album, bible, or torah.

"Wampum beads were the closest thing to written history for Wampanoag people," Paula Peters explained.

Wampum belts are often too large or beautiful to be worn on a regular basis. They have been used as invitations to formal events, to document and signify treaties, to declare alliances, and to commemorate family connections. Wampum adornments are not solely worn by tribal leaders. All tribal people could historically and may still wear wampum elements, like jewelry, often as a gift from a family member to bring luck and protection. The beads in various forms have found their way through networks of friendship, family, war, and trade far into the interior of the continent and overseas.

One of the most meaningful and earliest-described wampum belts, of the treaty type, is known today as "The Two Row Wampum," explained in the oral history and transcribed speeches of the Haudenosaunee (Iroquois). Their leadership had achieved remarkable peace among themselves, uniting multiple tribes, and in 1613 they negotiated a treaty with the Dutch, presenting these new settlers on the Hudson River with the Two Row Wampum to commemorate their agreement. On the belt, two purple-beaded bands ran parallel on a white-beaded river, one band representing the Dutch ship's path and the other a Native canoe's path, each riding parallel but on their own as equal, autonomous, and peaceful nations.

In the decades that followed, Dutch, English, and French colonists began to influence wampum as a trade commodity, which Native peoples welcomed as a way to communicate and trade. Archaeological records suggest that metal drills brought by the colonists enabled more of the working of the purple beads from the quahog, since the shells near the hinge are so thick. As the decades wore on, colonists worked with tribal communities or on their own to make inferior beads in small factories, using quahogs or other materials.

Quahog remains a prized food and fishery and form of artistic expression throughout the ancestral territory of the Wampanoag, Narragansett, and Mashantucket-Pequot tribes, among others. A Mohegan tribal leader gave a wampum pendant created by Elizabeth James-Perry, of the Aquinnah Wampanoag, as a gift to President Barack Obama.

In 2020, over a hundred people of the Mashpee Wampanoag tribe, led by Paula Peters and Linda Coombs, created a large ornate wampum belt of the community type, to tell a story, to revive the tradition of creating wampum within their community, and to serve as the leading object of a traveling exhibit in the United Kingdom that reflected on the four hundredth anniversary of the *Mayflower*'s arrival to Cape Cod. This wampum belt, made with nearly five thousand quahog beads, was also created to raise awareness about their search for an exceptionally significant belt, that of their ancestor Chief Metacom (King Philip), which was stolen from him after he was murdered in battle. It is possible the belt is still somewhere in England.

Wearing wampum earrings and a white and purple cotton sash around her waist that evoked the belts, Paula Peters ended a speech about the *Mayflower* story and English settlement in Wampanoag territory this way: "I do not hold you accountable for the actions of your ancestors: I hold you responsible for the future."

Right Whale

Although the right whale is less than half the size of the blue whale, the largest of the whales, a female right whale can grow to nearly sixty feet (18.3 m) long and weigh about seventy tons (63, 500 kg). That's the weight of nearly twenty-eight mid-sized trucks.

Biologists usually organize the right whales, blue whales, humpbacks, and about a dozen other species into a larger group known as the baleen whales. The other larger group of whales, which diverged over thirty-four million years ago, are the toothed whales, such as Risso's dolphins, killer whales, and sperm whales. Baleen is the filtering, fringy mouth plate material, consisting of keratin—the substance in horse hoofs and

North Atlantic right whale (*Eubalaena glacialis*) feeding on zooplankton with open mouth and baleen.

human fingernails. Baleen hangs down from a whale's upper jaws to sift out its diet of krill, small fish, copepods, amphipods, and other large aggregations of small marine organisms. Right whales are distinguished by their black, rotund body, large triangular flippers, and the absence of any dorsal fins.

On their enormous heads, right whales have thick patches of callosities, rough skin in the same places where we have hair on our faces: above and under their dramatically downturned lips and over their eyes. Barnacles and cyamids colonize right whale callosities a few months after the whales are born, creating a visual pattern that's unique to the individual right whale for its entire life, which might be a century or more—if left alone by people.

Early humans ate the meat (the muscles) and the thick blubber (fat) of any whale that beached itself ashore. Over millennia,

Whale barnacles (*Tubicinella major*), up to 1" wide (2.5 cm) and several inches deep into the skin, and cyamids, "whale lice" (*Cyamus ovalis*), both living on a right whale's callosities.

Indigenous peoples around the world began to spear whales while standing on flats of ice, sandy beaches, or rocky shores. People then used boats they covered with animal skins, carved paddles, and fashioned spears, harpoons, and nets to hunt whales from the water. Although this began perhaps far earlier in Korea or elsewhere, archaeologists believe the earliest documented people to go to sea and hunt whales seem to have been the Indigenous people of the Aleutian Islands at least fifteen hundred years ago. They hunted right whales.

Beginning over a thousand years ago, Norse and Europeans communities along the coast of the northwestern Atlantic caught right whales, too, boiling the blubber for oil to burn in lamps. In later centuries, people also used the oil to lubricate small machinery and used baleen as a strong flexible material for farm tools, whips, hoop skirts, umbrellas, and a range of other goods before the use of spring steel or the invention of plastics. Because of their slow nature and coastal habitat in temperate waters, similar to the great auk and Steller's sea cow, right whales suffered at the hands of humans more than any other type of great whale. "The North Atlantic right," explains expert David Laist, "was one of the first marine species to be pushed to the edge of extinction by human hands."

Right whales, in addition to being coastal and slow swimmers as compared to the other whales, are significantly more rotund. The name-derivation story says that these whales became known as the "right" ones to hunt, the correct ones to harpoon, because they were valuable and approachable. Hunters in boats powered by oar or sail just might have a chance of bringing these animals home. Right whales have also been known as black whales and Greenland whales. Only recently have biologists confidently differentiated between three geographically separate species of right whales and their close cousin, the bowhead whale (*Balaena mysticetus*) of the Arctic.

Two of the three species of right whales are now critically endangered. The North Atlantic right whale (*Eubaleana glacialis*) as of 2021 had fewer than 370 individuals remaining on Earth, even though humans haven't hunted them for a century. The North Atlantic right whale's recovery is severely hampered due to entanglements in fishing gear, strikes by large ships, the disruptions of ocean noise, and shifts in their diet due to climate change. The North Pacific right whale (*E. japonica*) is also in crisis, likely at a similarly low level, but there's far less scientific tracking of this species.

The southern right whale (*E. australis*), however, is doing far better today, which is a glimmer of good news. All of this requires a bit more historical background.

By the mid-1700s, whalers, hunting from sailing ships and rowing out with smaller boats and using harpoons, had largely stopped catching North Atlantic right whales. They could not find them anymore. So they turned to sperm whales in deeper water, another large whale species that they were slow enough to capture under sail or oar power—the kind of hunting done by Thomas Albro's shipmates. Commercial whalers in the 1800s would have been quite happy to kill a blue whale, for example, but those animals were far too fast for their methods and are rarely found in pods. So whalers from New England, the United Kingdom, and Europe sailed farther and farther away and for longer periods of time, traveling from coastline to coastline, serially depleting local populations of southern and North Pacific right whales and then moving on to another place. In the 1830s, while Albro, Darwin, and Bennett were sailing in the South Pacific, whalers killed well over fifty thousand southern right whales. This ended within a couple decades because it was no longer profitable due to the development of oil alternatives ashore, the loss of the whaling fleet during the Civil War, and because right whales had become so hard to find anywhere in the global ocean.

Consider, for example, a whaling voyage described by Frederick Olney, the third mate of the whaleship *Merrimac* out of New London, Connecticut. Olney, a veteran whaler, was exceptionally tall, identified on census documents as "mulatto," and he was recently married, a religious man, and a teetotaler. Olney's ship left in the summer of 1844, but made no attempt to head along the coast for right whales. They had no hope of seeing any. The *Merrimac* immediately sailed for the Azores to follow the currents and winds to go south. Along their way around the world, past the Cape of Good Hope and across the Indian Ocean, they saw "grampuses," a few sperm whales, and some finback whales—but no right whales.

As Olney's ship neared the coast of Australia, they captured a sperm whale. Then a day or two later on December 1, 1844, Olney wrote in his journal: "Thursday last was supposed to be Thanksgiving day at home. And our share of their annual feast consisted in the reflections of the occasion. But the day did bring forth something unusual for us. For at early dawn, say not past 4 o'clock, we saw 2 right whales about 2 miles distant. A sight so rare that it really caused a spontaneous burst of thanks."

They lowered their boats and managed to plunge harpoons into two southern right whales. One whale got away—they had to cut him free because the lines got tangled—and the second they killed, "but to our mortification," Olney wrote, "it sank."

They returned to the ship empty-handed.

Olney wrote that by 6 p.m. they had hauled the boats back up, finishing his entry: "and we on our course East without our Thanksgiving dinner in cutting a whale, . . . I turned in to dream of my [wife] Olive and pumpkin pies."

Some derivations explain that these were also called "right" whales because they floated when killed, making them easier to butcher, but this was not always the case—as Olney and his shipmates proved here.

As Olney was sailing in the 1840s, whalers were rapidly depleting right whale populations from the coast of Australia and Aotearoa New Zealand. On this voyage, Olney's captain barely bothered with this area and kept on going, navigating their ship up to the far northwest Pacific, off the Kamchatka Peninsula, which was at that point a fairly new commercial hunting area for right whales. Olney's *Merrimac* returned to New England after a nearly three-year voyage, filled with 2,975 barrels of whale oil and 5,000 pounds of baleen from North Pacific right whales, which represented a catch of maybe thirty to forty individual animals. This seems to have been a below average voyage, but not a financial flop.

Most whalers in the mid-1800s believed, or at least rationalized, that right whales were just swimming away from the hunters, hiding in the wide ocean like deer in a different part of the forest. A few scattered people within the industry, however, were sounding the alarm that they were not just chasing the right whale schools around—they were eradicating them. Right whales do live globally like sperm whales or blue whales, but live and migrate in fairly small regions of the sea. In 1845 a whaler named M. E. Bowles, for example, sent an account of hunting whales in the far northwest Pacific to a Hawaiian newspaper called *The Polynesian*. Bowles published a rare recognition of the systematic depletion of the right whale from region to region, for which the American whalers were largely responsible. Bowles thought the northwest Pacific was big enough that it might sustain the hunt for a while, but in the end, he concluded, "The poor whale is doomed to utter extermination; or at least, so near to it that too few will remain to tempt the cupidity of man, I have not a doubt."

As the hunting progressed, whalers did indeed "fish out" the right whales and the closely related bowhead whales off Kamchatka and sailed farther north into the Arctic of the North Pacific. Olney did this on his next voyage out of New London, with another captain and crew who focused their efforts on bowhead whales alone. By 1859, Mary Chipman Lawrence, sailing with her husband and daughter marveling at Mother's Carey's chickens, wrote when she spotted a right whale in the far North Pacific: "In the afternoon raised, for a wonder, a right whale. . . . It is encouraging to see one occasionally even if we are not to have them. We began to fear that they were all dead."

By the time large-scale industrial whaling began in the twentieth century—using engines, steel hulls, power winches, stern ramps, explosive harpoons, and other technologies beyond anything Olney, Bowles, Brewster, or Thomas Albro could envision—whalers were targeting blue whales and other large baleen whales. They hunted the whales for margarine, soap, meat for humans, feed for livestock, and a range of other products. Right whales and sperm whales were captured only on days when they couldn't find any other baleen whales. Right whales had become an afterthought.

Aside from a few exceptions, commercial whaling has largely stopped thanks to an international agreement of a moratorium beginning in 1985. There's reason for optimism, as it appears many whale populations are recovering, if slowly. The southern right whale, for example, which once likely had a population of fewer than 500 individuals in the 1920s, nearly as low as its northern cousins, has been coming back strongly due to protections and coastal habitats that are less populated by people, fishing gear, and large ships. Scientists believe that as of 2021, there are about fifteen thousand southern right whales, with the populations around Brazil, South Africa, Australia, and Aotearoa New Zealand recovering especially well, growing at a rate of 7 percent per year and repopulating former regions where they had not visited in nearly a century.

The southern right whale is on the right track, revealing that there is still hope for the two species in the Northern Hemisphere if we act immediately.

Sea Cow

The *St. Peter* gropes somewhere among the long desolate string of Aleutian Islands in the far North Pacific on their way home. They are a Russian discovery ship, westbound, returning from a trip to explore the coast of what is now called Alaska. They've long since lost track of their sister ship, the *St. Paul*. It is September 1741. The situation is dire.

All the sailors aboard the *St. Peter* recognize it is unlikely they will ever make it home. Scurvy has flattened several of the crew. Two men have already died. Captain-commander Vitus Bering himself is terribly ill, perhaps with scurvy and heart disease. The fresh water in barrels is mostly foul. Storm-force winds and crashing seas are relentlessly in the wrong direction.

Sharing the cabin with Bering is a German physician and naturalist named Georg Wilhelm Steller. Although he's on his first voyage, Steller is certain they are near to land because he sees floating seaweed and various birds that he knows to be strictly coastal. No one listens to him, in part because he's a pompous know-it-all, and, in this case, because he's also wrong about where they are exactly.

Bering's expedition continues clawing westward toward Siberia's Kamchatka Peninsula. Gales rage on. More men die. Eventually, somehow, in the middle of the night they stumble over a reef, barely making a small protected harbor. They drop the anchor and sleep, praying this is the mainland.

In the morning, Steller and his servant row several of the sickest men ashore. Steller thinks that the place must be an island because of the shape of the clouds and how the sea otters carelessly swim over to the boat, unafraid of humans.

Once ashore, Steller notices a mammoth animal swimming along the coast. Neither whale nor shark, it is a creature that he has never seen before. As far as he knows it is entirely unknown to the scientific community in Russia and Europe.

Steller writes in his journal, later translated from German: "Nor could I even know what kind of an animal it was since half of it was constantly under water."

As the shipwrecked party settles into a camp to winter on the island, they remain wracked with scurvy and hunger. They are tortured by the Arctic foxes that steal food from the camp and bite at the sick men. More of the sailors die. Yet as Steller provides antiscorbutic plants to eat, others begin to recover. But then a storm drives their ship aground, thumping an irreparable hole in the hull. Commander Vitus Bering dies. They bury him in a frozen shallow grave. The island now carries his name, but it

The extinct spectacled, or Pallas's, cormorant (*Urile perspicillatus*) fishing underwater.

will take the crew months to confirm this is an island for sure, because it's fifty miles (80.5 km) long.

With Steller and the first mate bickering, the remaining sailors, weak and freezing, struggle through the long winter snow and blizzards. They dig out makeshift shelters and eat spoiled ship's flour, wild birds, foxes, sea otters, seals, and wild plants. Even after they have eaten through all the animals and greenery they can reach, they are still starving. This includes hunting and eating enormous twelve to fourteen-pound, likely flightless seabirds. These were spectacled cormorants (*Phalacrocorax perspicillatus*), closely related to the guanay cormorant, and they would soon go extinct due to hunters without the rock-star publicity of the dodo or the great auk.

After the winter ice has melted, the men of the *St. Peter* turn ravenously, drooling, to those strange gigantic animals they have watched from the shore. They repair one of their small boats and fashion a sort of harpoon. After many failed attempts, the men are able to kill one. It is a sort of manatee, what would later be known as a sea cow, a female, which they drag to shore.

"At long last," Steller writes in his journal, "we found ourselves suddenly spared all trouble about food."

The naturalist explains that these sea cows are completely unafraid of people.

"They have indeed an extraordinary love for one another, which extends so far that when one of them was cut into, all the others were intent on rescuing it and keeping it from being pulled ashore by closing a circle around it. Others tried to overturn the yawl [boat]. Some placed themselves on the rope or tried to draw the harpoon out of its body, in which indeed they were successful several times."

For the next few days, a male sea cow keeps swimming by, presumably to watch his dead mate on the beach. The men feed on the meat. They use the oil and the skin from the animal. Now forty-six survivors, they are able to live long enough to finish building a new, moderately seaworthy boat to sail out beyond the island's southeast point, which they have named Cape Manati, after the sea cows, and then west across the final stretch of the North Pacific to the Kamchatka mainland and a path, by land, to their homes.

It was not long before their reports of the plentiful sea otters, fur seals, sea lions, and sea cows around Bering Island lured Russian fur traders and hunters who quickly expanded their trade to the North American coast. These men decimated the already small population of sea cows. Steller's detailed descriptions are the only account left of this animal alive, which is now known today as the Steller's sea cow (*Hydrodamalis gigas*). Aleutian and other Indigenous peoples had likely already hunted this animal to near extinction throughout much of its range, before the arrival of the Bering Expedition. The overhunting of sea otters,

Cartoon representation of the extinct sea cow (*Hydrodamalis gigas*), of which there are no known detailed illustrations or photographs; here in comparison to the dugong (*Dugong dugon*), which at ~10' long (3 m) would be a third the size of the sea cow.

some scientists think, might also have harmed sea cows, since without sea otter predation, urchins can decimate kelp beds, the food of the sea cows. Regardless, once the Russian hunters got involved, the sea cows, living in these last few places, were wiped off the Earth in less than thirty years. Little is left today beyond stories, fossils, bones, a cartoony sketch from one of the officers of the Bering expedition, and Steller's observations and the measurements from his dissections.

Steller's sea cows grew to be nearly thirty feet long (9 m) and weighed perhaps nearly thirty thousand pounds (5,900 kg), which is just about the length and weight of a mid-sized American school bus. Sea cows had proportionally small heads, small eyes, and a stiff, whale-like tail. To chew kelp and other algae, since they were vegetarian grazers, they did not have teeth but two upper and lower molar-like plates with ridges to grind the vegetable matter. The sea cow had a stomach larger than a sleeping bag.

"It is covered with a thick hide," Steller wrote, "more like unto the bark of an ancient oak than unto the skin of an animal; the manatee's hide is black, mangy, wrinkled, rough, hard, and tough; it is void of hairs, and almost impervious to an ax or to the point of a hook."

The "strangest" characteristic, Steller explained, was the stubby, arm-like flippers that helped the sea cows swim, walk in the shallows, pull out grasses, and hold each other while mating. He described this sea cow appendage, which had no claws, as like a covered hoof, like "an amputated human limb . . . covered with skin."

Although some of the physical traits are like those of a walrus, the closest living relative to the Steller's sea cow is the dugong (*Dugong dugon*), which lives in warm coastal waters around the Indian Ocean, Australia, and the southwestern Pacific. The Steller's sea cow and this living dugong are also close relatives, within the order Sirenia, of the three species of manatees (*Trichechus* spp.), which are found in tropical fresh and salt waters around the world. Most scientists now believe the sea cows, manatees, and dugongs are more closely related to elephants than they are to the walruses or grampuses. If the longevity of the Steller's sea cow is anything like their living relatives, individual sea cows might have lived in the North Pacific to be over sixty or seventy years old.

For his part, Georg Wilhelm Steller died soon after he returned to Siberia, at only thirty-seven. He did not survive to report his discoveries in person. His shipmates, wanting every spare inch of space for sea otter pelts, had not allowed him room in the boat to bring back a full skin of the sea cow. In his account, published in 1751 after he died, Steller wrote, "No one who has studied various lands doubts that the vast ocean contains many animals which today are unknown."

We can confidently say the same today.

Sea Pickle

Tonight it is Ella Cedarholm, a young oceanographer, who is the assistant scientist in charge of processing the net tow. It is 1:15 in the morning. The breeze is cool but not freezing, so she needs only a light hoodie sweatshirt, especially since she is walking back and forth between the deck on the port side and the small laboratory inside.

As Cedarholm organizes the college students as to what to separate, analyze, and quantify, she stops in the wet lab and looks in on one bucket with a few odd-looking large objects floating in the water. The movement glitters its shape with flickering bioluminescence. She turns on her headlamp for a better look. Now in red light, she pokes one with a spoon. Then she picks it up with her bare hands, and stands up, leaning her hip on the edge of the steel sink because the ship is heeling over to starboard. The thing in her hand is a rubbery, nubby clear something, like a toddler's play rolling pin, but more springy and bendy, kind of like—a huge pickle?

This is the austral summer, February 27, 2020. Cedarholm and the rest of the crew of the oceanographic tall ship *Robert C. Seamans* have been sailing about a hundred miles off North Cape, the northernmost tip of Aotearoa New Zealand.

While Cedarholm and her student group were sleeping, the previous watch had deployed a long, fine-meshed net with a metal ring that has an opening that is one meter wide. They had sent this net down several hundred feet beneath the surface, towing it as the ship sailed very slowly along. When they hauled the net back aboard, using a power winch, they had found in the back of the net a luminous greenish clump of these rubbery thingies. This group of students was near the end of their watch; they still had to set more sail and clean up the lab before they could go to sleep below, so they had left the catch in the bucket in a secure place for Cedarholm and her group to examine during their watch. This was common protocol.

Suddenly Cedarholm realizes what's in her hand. It's a pyrosome.

She keeps the animal lit with her headlamp, switching over to a bright white. She gathers the students around. They are at once amazed but also elbowing each other and giggly because, illuminated this way, it has a comic resemblance to some kind of sex toy. Cedarholm ignores the smirks and explains that this pyrosome is a set of living colonial organisms. Each of these cylinders is encircled with nubby spikes. And each nubby spike is its own animal, called a *zooid*, with a simple brain, a stomach, a liver, and a basket with cilia that acts as a filter. All of these zooids, each a clone of its neighbor, are connected to the others with a clear gelatinous material that forms the firm sock-shaped colony that opens on one end.

"Awesome," says one of the students, and means it.

Pyrosomes, it turns out, have inspired an unexpected level of poetic enthusiasm among historical seafarers. The first Western author to describe and formally name these "sea pickles" or "sea cucumbers," as some sailors appropriately called them (not to be confused with *bêche de mer*, the echinoderms, also known as sea cucumbers), was the French explorer and naturalist, François Auguste Péron, who in the early 1800s encountered a large patch of pyrosomes one night at the surface in the equatorial Atlantic.

"Ce spectacle," wrote Péron, "au milieu des circonstances que

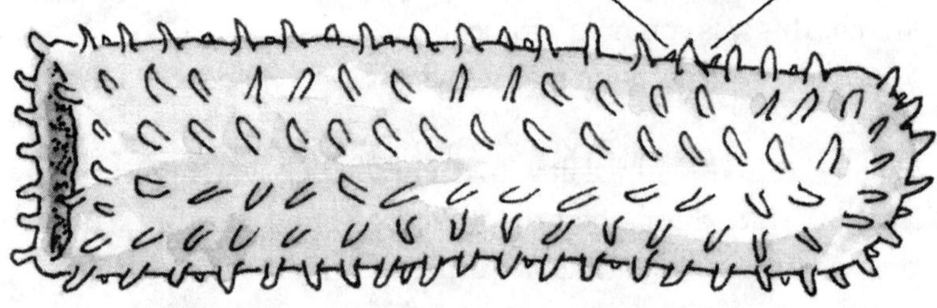

Pyrosome (*Pyrosoma* sp.) with inset of a zooid: black dots are the bioluminescent organs above the filtering basket and stomach.

je viens d'exposer, avoit quelque chose de romanesque, d'imposant et de majestueux qui fixa tous les regards," which has been translated as follows: "Heightened by the surrounding circumstances, the effect of this spectacle was romantic, imposing, sublime, riveting the attention of all on board."

Péron brought several specimens on deck, all between three to seven inches (7.6–17.8 cm) long. He described their anatomy. He tested their bioluminescent properties. And it was he who gave them their scientific genus name *Pyrosoma*, which means "fire body" in Greek.

In the decades to follow, other seagoing naturalists wrote about pyrosomes with similar poetic awe. Francis Allyn Olmsted, a Yale undergraduate who went to sea on a whaleship for the sake of his health in 1840, said that in big seas the jelly-like creatures rolled around the ship's deck like "a fire-ball." A few years later in 1849, English biologist T. H. Huxley wrote in his diary of the "beautiful Pyrosoma" that were "shining like white-hot cylinders in the water."

It was Dr. Frederick Bennett, however, the same ship's surgeon who carefully analyzed the pilot fish, who left one of the most passionate published descriptions of pyrosomes, as if to better even François Péron. Bennett, sailing as a surgeon aboard a whaleship in the 1830s, was not prone to flowery exaggeration. Yet he could not contain himself when it came to writing about pyrosomes. Like Péron, Bennett observed them in enormous numbers in the equatorial Atlantic. The pyrosomes were in "their zenith of splendor," he wrote, aggregating in such densities that the whalers could fill any number of buckets with them. The mass of pyrosomes created such a glow in the sea that the animals rendered, Bennett wrote, "all surrounding objects visible during the darkest night, and [permitted] a book to be read on the deck, or near the stern-cabin windows of a ship."

Marine biologists know today that these colonial organisms float freely at sea, anywhere from on the surface to as far down as three thousand feet (915 m). They are filtering organisms. The zooids unite in a common body to swim around, in their fashion, and thus have better success at collectively capturing phytoplankton, the microscopic plants they like to eat. Each zooid excretes water and waste out their shared opening into the middle of the pyrosome. This collective filtering and ejection helps them move their body along, allowing them to control a bit of their vertical motion and even achieve some intentional propulsion.

Cedarholm leads her students into the dry lab. They examine the pyrosomes with hand lenses and microscopes. Each individual zooid of a pyrosome has what scientists believe to be cells with bioluminescent bacteria. The pyrosome's famous bioluminescence has been recorded as a range of blue, green, and even pink. The glow helps them attract phytoplankton toward their collective body, as well as perhaps to communicate with their fellow zooids so they might collectively adjust their pyrosome's vertical motion, up or down, following their food as it moves up or down in the water column in response to sunlight. This might also help them escape becoming a snack for their predators, such as sea turtles and small whales, grampuses and dolphins.

Unfortunately, Cedarholm and her students can't spend their entire watch looking at the pyrosomes. There's a lot of other work to do—copepods to count and microplastic fragments to filter—so she and her watch throw overboard all the pyrosomes except for one of the largest, leaving it slopping around in a spare bucket so students of the morning watch can take a look.

By 0930, the sun is up and strong, and Cedarholm's watch is now asleep. Sloshing in its lonely bucket, the pyrosome is no longer transparent, but turned orange and become more rigid, like a large nubby chew treat for dogs.

These waters north of Aotearoa New Zealand are famous for exceptionally large pyrosomes. The longest Cedarholm measured was about one foot long (.3 m), but her chief scientist on the voyage had one night sailed over a different type of pyrosome in these waters that appeared larger than bioluminescent boulders. They had thought at first that they were about to run aground. Since Péron's day, marine biologists have identified several different species of pyrosomes, which include some individuals that grow to seriously enormous lengths. In 2018, for example, in waters a few hundred nautical miles to the south of where Cedarholm and the *Robert C. Seamans* sailed that night, two scuba divers happened upon a giant pyrosome that was over twenty-five feet (7.5 m) long. In 1969, even closer to the North Cape, biologists found a pyrosome that was about thirty feet (9 m) long, with a diameter of three feet (.9 m). That's enough to make a sailor run out of glowing terms.

Silver King

A twelve-year-old boy named Harry Dean lived in Philadelphia in 1876. His uncle, Captain Silas Dean, came for a visit and persuaded his parents to let him bring Harry along for a three-year trading voyage around the world. Harry was desperate to go, so his parents reluctantly agreed.

Off went Harry aboard his uncle's ship, *Traveller the Second.* He was part of a long line of Black seafarers and merchants. Harry's mother, Susan Cuffee, was descended from the famous Captain Paul Cuffee Sr., whose parents were African American and Native American of the Wampanoag and Pequot tribes. Paul Cuffee had built and owned his own first ship, the *Traveller*, and he'd skippered all-Black crews and became a luminary for his community. Silas Dean, for his part, was a tall, powerful man who was very good to his nephew, but out at sea as captain his uncle spoke little beyond his booming commands on deck to his crew.

They sailed south along the US East Coast and then steered *Traveller the Second* toward the Straits of Florida. Harry acclimated to life at sea. He surely learned how to pause at the right moment when walking up a ladder when the boat was heeling over on a wave, learned how to do a trick at the wheel, and reveled in the soft pink sunsets of a full ocean horizon at the end of a day at sea.

They entered the Gulf of Mexico, and, as he told the story years later, Harry went back to the quarterdeck to try to catch a fish. The mate, who also loved to fish and had his own bait in the water, made fun of him, probably because Harry streamed a line over the side that was extra thick or maybe he had rigged an enormous hook.

The mate said, "Sonny, what you fishin' for—whale?"

Yet soon a bite on his line nearly wrenched Harry right off the ship. He braced himself against the rail. He shouted for help. The mate and a couple other sailors rushed over, all working together to haul it in.

"As they pulled a great fish jumped out of the water," Harry yarned. "The sun struck his scales, making a gleaming silver halo about him."

"Shiver my timbers if he ain't a silver king!" said the mate, who was now both jealous and impressed. The men managed to bring the fish aboard. It weighed just over seventy-five pounds (34 kg).

Harry declared, "It was the most delicious fish I have ever tasted."

Still in use today, "silver king" is another name for the giant tarpon (*Megalops atlanticus*), a fish that is known to glisten with an especially silvery sheen due in part to its large scales. Giant tarpon are found only in the Atlantic Ocean, but they range from as far north as the Canadian Maritimes to as far south as Argentinian waters, and from as far west as the Caribbean Sea across to the coasts of Africa, from Mauritania down to Angolan waters.

They don't call them giant tarpon for nothing. These fish can weigh over three hundred and fifty pounds (160 kg). The females grow larger, estimated at longer than 8.5 feet (2.6 m). In 1991, a fisher off Sierra Leone caught a silver king that was 283 pounds (128.4 kg), which actually, quite curiously, tied an earlier world record at the time from the 1950s, when another fisher caught a silver king in a lake in Venezuela that also weighed 283 pounds.

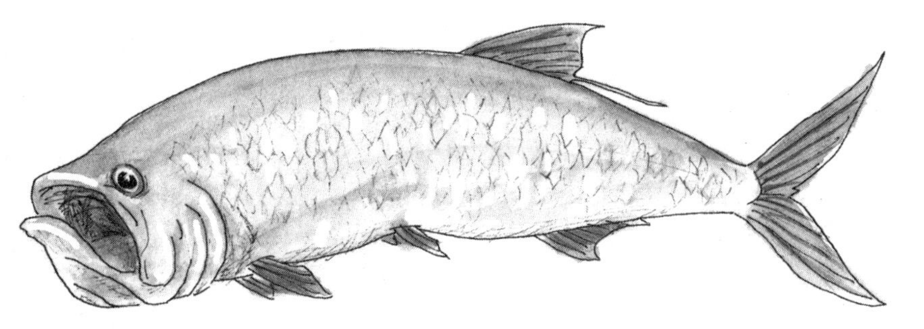

Silver king, or the giant tarpon (*Megalops atlanticus*).

The heaviest silver king caught on record in American waters was 202.5 pounds (91.9 kg), hauled in near the coast of Florida in 2001, not far from where Harry Dean caught his tarpon and where Hemingway's Santiago killed his marlin.

Silver king has been a popular species for recreational fishers since at least the mid-1800s. Martha Field, when reporting of the shrimp fishery, also wrote of the appeal of casting for silver king and the abundance of this fish off coastal Louisiana.

Today, despite Harry's endorsement, giant tarpon is not popular for eating in the United States, but it still is in in parts of the Caribbean and in Africa. In US waters the tarpon fishery is closely regulated, and you may only catch-and-release. Yet still tarpon appear to be in decline globally, officially classified as vulnerable by the IUCN, likely because of overfishing and habitat loss.

One of the silver king's special characteristics is that, although it likes warm water, this normally pelagic fish is especially tolerant of different salinities and oxygen levels. Tarpon can visit freshwater environments for parts of their life, which is how a giant tarpon heavier than LeBron James was found in a lake in Venezuela. When tarpon are in an environment that doesn't provide enough oxygen, they have a special ability to gulp air at the surface. Young Harry would've noticed when he brought up his catch that tarpon have a huge, wide, gaping mouth. The fish use this enormous mouth for suction: they literally inhale crab, shrimp, polychaetes, and smaller fish, such as mullet, sardines, and needlefish. Their large, mirror-like scales, which can grow as large as a human's palm, help the fish blend into their underwater surroundings. Tarpon grow and mature slowly, looking to escape predators like sharks, alligators, and porpoises. A few tarpon in captivity have lived more than sixty years.

The silver king that young Harry brought aboard on his first voyage would hardly be his last wonder of the sea, let alone the last fish he would catch from over a ship's rail. After returning to Philadelphia at the end of his three-year adventure sailing around the world with his Uncle Silas, Harry Dean went on to become a ship captain himself, as well as a merchant, a shipowner, and a political activist in Africa and back home in the United States. Dean did a bit of sealing, too, but found that he could not stomach the cruelty of the business. "There was money in it, no doubt," Dean said, "but I found little excitement in clubbing to death beasts as mild-natured as dogs."

When he was in his sixties Dean relayed his memoir to a historian, published as *Umbala* or *The Pedro Gorino: The Adventures of a Negro Sea Captain* (1929). It was in this book that he spun that tale about the day years ago that he remembered so well, how he caught his first big fish, a silver king in the Gulf of Mexico.

Teredo Shipworm

Now if you think that a long, slimy ocean invertebrate is an inconsequential animal, then consider the declaration from the famed ecologist and environmental historian James T. Carlton: "Shipworms were one of the greatest forces controlling the history of global shipping, regulating the evolution, survival, and routes of wooden ships for thousands of years."

Here's how it happens: floating in the ocean, a tiny teredo larva lands on a hull or piling or piece of driftwood and immediately begins to grind, to bore into the surface of the wood with two rasp-like encircling shells. Shipworms are not true worms, but instead a type of bivalve clam, related to quahogs, but with a thin, soft, worm-like body and a relatively tiny shell. With the

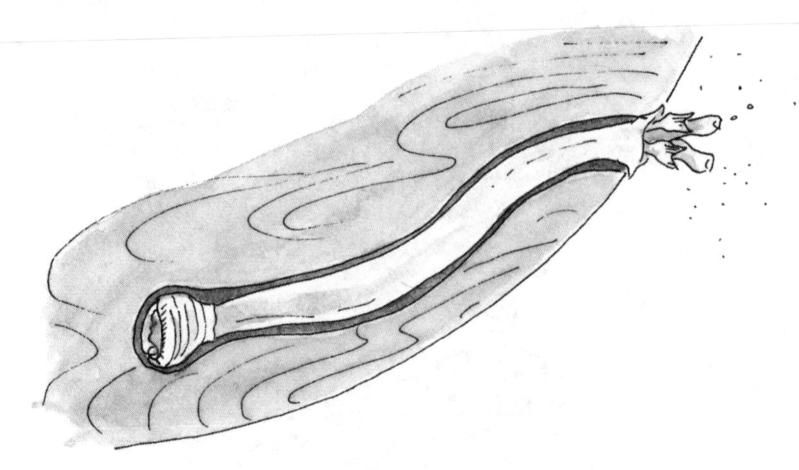

Shipworm clam (*Teredo* sp.) in a cross-section of a wood hull.

help of internal bacteria, the teredo clam gets the nutrients it needs from the wood. With a two-part siphon as a sort of tail, it sucks in water from one opening and expels its waste with the second. Shipworms have an internal gill that runs nearly the length of their bodies. With a shell plate at its tail end, it can bring in the two siphons and close the back door, where the hull meets the water, so it won't be disturbed by nibbling fish or if the seawater environment is not to its liking, such as if the water is too cold or too fresh. In this way, shipworms are able to live for weeks in some woods without air or water.

Not much is funny about the disastrous impact of European first contact on the Indigenous peoples of the Caribbean and the Americas, but it remains a rather humorous historical footnote that Columbus's final voyage, after most of his crew survived a rebellion from a tribe of Mayan warriors, was finished in the end by a slow-motion welcome from an invasion of shipworms.

Columbus led four separate voyages from Spain to the Caribbean, during which he observed frigatebirds, dizzying populations of sea turtles, and individuals of the now extinct Caribbean monk seal. In his last voyage, after tracing the eastern coast of Central America, having to abandon one of his four ships, and mourning the death of several of his sailors, including some of the caulkers, Columbus decided to cut his losses in April of 1503 and bring back his lusty reports of gold and silver to the monarchy of Spain. Yet as he was retreating from his final anchorage, which was off present-day Panama, Columbus noticed something terrible. His vessels were so compromised by shipworms that they were barely seaworthy. With little choice, even knowing his three remaining ships were in bad shape, Columbus sailed off to try to catch the trade winds home—or at least limp over to the more friendly Spanish settlement on the coast of the present-day Dominican Republic.

Before long, though, Columbus had to abandon another ship of the fleet because of these teredo worms. He piled all of his men into the two remaining caravels. The pair of ships tacked their way up into the Caribbean Sea to some islands to the south of Cuba, which he named "Las Tortugas" because of the abundant sea turtles there (these islands would later be named the Cayman Islands, the home to Captain Bush). Stormy weather and trade winds spoiled any real progress eastward. They lost most of their anchors. But it was the shipworms in the end that were the voyage's greatest hazard.

Columbus wrote in a letter to Queen Isabella and King Ferdinand: "My ships were pierced by borers more than a honeycomb, and the crew entirely paralyzed with fear and in despair." He rested in a safe harbor for a week, then tried to go sailing once more, ordering his crew to pump the bilges for dear life.

"I put to sea again," he wrote in the letter, "and reached Jamaica by the end of June; but always beating against contrary winds, and with the ships in the worst possible condition. With three pumps, and the use of pots and kettles, we could scarcely clear the water that came into the ship, there being no remedy but this for the mischief done by the ship-worm."

Now desperate, Columbus was forced to beach his two sinking ships in what is now St. Ann's Bay on the northern coast of Jamaica. Here, shipwrecked with about one hundred men, he wrote the letter to Ferdinand and Isabella to explain the wormy situation and why he would be asking at Hispaniola for ships to rescue him and his crew.

Many species of shipworms, these long-bodied clams with their pair of shells, live all around the world. One of the most common shipworms in the North Atlantic that bore into wooden ships, pilings, docks, and dikes has the scientific name *Teredo navalis*. This comes from the Greek for "wood worm" and from the Latin "of ships." The shipworm species that ate up Columbus's fleet was one of several different, larger species that are more common to tropical waters, some of which can grow up to two feet (.6 m) long. (There are tropical species in the Pacific that can grow six feet, 1.8 m, long.) Regardless of species, however, mariners for centuries have also called the shipworms, in general, teredoes.

Shipworms tunnel parallel to each other. They communicate in some fashion, somehow, so in the hulls of Columbus's ships they sensed each other's path, never crossing, thus slowly eating away the planks from the inside.

As early as 350 BCE, Theophrastus had written about shipworms compromising the vessels of ancient Greece. The clams continued to plague mariners' ships until the twentieth-century invention of steel and fiberglass hulls. (I've not yet been able to find any records or oral histories of shipworms affecting the early Polynesian voyagers, but they must have had to contend with them, too, unless they found tropical woods resistant to this kind of fouling.) As noted earlier, Alexander Selkirk, the real-life model for Robinson Crusoe, did not maroon himself in the Pacific in the early 1700s because of his love for Juan Fernández crawfish, but because his ship was sinking due to shipworms. Professor Carlton explains that ships returning to Europe in the 1730s likely brought back the cold-water *Teredo navalis* from the Pacific Ocean, releasing a plague of shipworms that crippled the entire wooden dike system of the Netherlands.

Early mariners tried to use coatings, such as wax or tar, on their ship hulls to keep the shipworms out. Columbus tried a similar strategy, using tallow and pitch. Some early navigators tried to spend time in freshwater harbors or rivers to try to kill the clams. Columbus tried this too, but not likely on purpose—and it didn't work for him anyway. Even today, though, if wooden pilings and docks are not regularly treated with chemicals, shipworms can be a problem. This is often an unexpected and ironic result when we remove toxins and improve the water quality in a harbor and waterway—our wooden docks begin to crumble.

The ancient Greeks tried to barricade their hulls with copper sheets. This deterrent was revived in the 1700s and 1800s with copper sheathing tacked to wooden hulls, as well as later bottom paints with copper and other chemicals that are poisonous to shipworms and other fouling organisms, such as gribbles and barnacles. These strategies could be effective. For example, historians explain that the copper sheathing used on English ships made the difference in a major battle in 1780 against a Spanish navy whose vessels were weakened and already sinking from shipworm holes. Shipbuilders of large wooden vessels also took to fastening a "worm shoe," which is a layer of sacrificial wood beneath the keel.

Back in Jamaica in 1503, with his ships irreparable from teredo damage, Christopher Columbus and his crew remained on the beach. They survived their own mutiny, lack of food, conflicts with Taino peoples, and just had a generally miserable and destructive time of it. Finally in June of 1504, a year later, two Spanish ships from their settlement on Hispaniola arrived to rescue them. Columbus abandoned the deteriorated hulls and rode a new ship back home to Spain. This time Columbus stayed in Europe for good and spent his time retired from the sea, surely spinning yarns about the terrible teredo worms, about which he would not clam up.

'It's not that I've lost interest in the New World,' I like to imagine him saying, 'it's just that at sea I was nearly bored to death.'

Tropicbird

By December 15, 1952, Ann Davison had been at sea alone for over three weeks. Aboard her twenty-three-foot sailboat, *Felicity Ann*, she was still only a third of the way through her ocean crossing from the Canary Islands to the Caribbean, which would be the first known solo ocean crossing by a woman. Davison had yet to find the favorable trade winds to send her west, and there was no way to contact anybody for weather reports. At that time there was no such thing as GPS, satellite phones, or long-range marine weather broadcasts for small boats on the open sea. So all this English sailor could do was to keep calm and carry on. Davison tried to locate the normally predictable trade winds upon which sailors usually depend in these latitudes, the same which Columbus had identified and introduced to Europe, yet these trade winds were inexplicably nowhere to be found.

Around noon of that day Davison observed "enormous shoals" of flying fish with dolphinfish splashing in pursuit. Large flocks of seabirds soared and dove into the ruckus. Within this feeding frenzy, the water's surface furious and frantic with marine life, one particular pair of birds caught her attention.

"Two bosun birds visit ship with loud squawks and fly around and around mast shrieking ornithological oaths at one another," she wrote in her journal. "Flying appears to be a great effort to them and they seem to be on the point of stall unless flapping furiously with their narrow swept back wings."

Davison's birds were tropicbirds (*Phaethon* spp.), a small

White-tailed tropicbirds (*Phaethon lepturus*).

group of seabirds well-known for circling around boats at sea and occasionally resting in the rigging. Based on her geographical location, the two flying above *Felicity Ann* were likely of the red-billed or white-tailed species of tropicbirds. They were probably coming in from the Caribbean or Bermuda, where they have scattered breeding colonies, but they could also have flown from the Canary Islands, Cape Verdes, or from other island rookeries in the eastern Atlantic. Tropicbirds have glistening white plumage as adults, and their narrow central tail feathers flow behind them like streamers. These tail feathers, found on males and females, can grow longer than their bodies, nearly as long as the bow of a violin. Sailors historically called them bosun birds, probably because their tail feathers evoked to them a marlinspike, the long thin metal tool commonly worn on the belt of a ship's bosun (from boatswain), whose job it is to maintain a ship's sails and rigging. The squawks and shrieking of tropicbirds might account for another reason sailors called them bosun birds. On several types of ships, especially in the navy, the bosun carries a metal whistle to alert or deliver shrill commands that their crews can hear over the wind and waves.

Although Davison did not aspire to be an ornithologist, she was a careful writer and observer. She made accurate diagnostic observations of her tropicbirds, which are indeed known to hover with rapid wing-beats, pigeon-like, a unique flying style among deep ocean seabirds. Tropicbirds, true to their name, are found mostly in warmer waters. They soar sometimes higher than frigatebirds and often plummet straight down from great heights, plunge-diving for fish or squids. Like other plunge-divers, such as pelicans, tropicbirds have air sacs within their head and neck to cushion against the impact of the water. At other times, more like noddies, tropicbirds snatch prey just off the surface, such as flying fish and needlefish trying to escape other predators in the water, as they were doing during the feeding frenzy Davison witnessed in the middle of the Atlantic. Tropicbird beaks are serrated to allow them a good grip on their prey, They are also some of the more far-ranging of the seabirds, traveling thousands of miles from their breeding sites, with no predictable migrations that ornithologists have yet understood.

Four days after her first visit from that pair of tropicbirds, Davison was sailing slowly along and saw another one of these birds from the cockpit of her boat.

"Bosun bird, going east," she wrote in her journal. "Flies past ship without batting an eyelid. This very unusual as these birds take a great interest in *F[elicity] A[nn]*. So I give a piercing wolf whistle, whereon bosun bird wheels smartly around, flaps back to ship, dives and stalls about three feet from my head, looks sharply at me, decides against whatever he had in mind to do and flies smartly away again."

Ann Davison went no further into this meeting in her published writing, in the book that would become a classic of sea literature, titled *My Ship Is So Small* (1956). Davison was not inclined toward lengthy, emotional responses to marine life of any kind in her prose, yet albeit brief, this was a profound moment: a meaningful meeting at the core of this ocean bestiary collection of little stories, because here was a person who had been alone at sea in a tiny boat by herself for nearly a month. Not quite in as dire a situation as Poon Lim with his flying fish, Davison was still terribly anxious as to where she was exactly and when the trade winds would fill in and, frankly, as to her very survival. She knew that she was at least several hundred miles and weeks of travel from any land, floating in a small wooden vessel in an environment in which humans as individuals are in no way equipped to live for any length of time. Now here she was, at her urging from her whistle that tried to speak another language, face to face with a representative from another species who very much did belong here, another animal fluttering above her that was fully comfortable living and navigating across this open ocean. Tropicbirds spend the majority of their life out at sea, usually alone. They have large black eyes which blend into a thick streak of black plumage that grows across and below them, presumably evolved to help with glare while spotting prey below. December can be a time for breeding for tropicbirds. Maybe this was a female tropicbird, out foraging while her mate sat on their single egg. Whoever this wild bird was, whatever its thoughts, its history, its destination, this one tropicbird responded to the human sound, had paused to flutter and look down on the other on the floating boat. Here was an increasingly rare moment out in the deep ocean environment in which a human and another animal paused to consider one another—eye to eye.

However Davison truly processed that moment with the

tropicbird, she continued on her crossing, across an ocean she described as an inhospitable, callous place that offered only personal challenge. Davison's Atlantic was not a setting of partnership, sustenance, nor one that needed human stewardship. After an exhausting nine-week passage in which she never found the trade winds, Ann Davison at last completed her history-making, single-handed trans-Atlantic sail when she dropped anchor in a quiet harbor of Dominica just before sunset on January 24, 1953.

She rested up, reconnected with her people, then sailed *Felicity Ann* alone up the Windward Island chain to Antigua, on to the Bahamas, and finally up to New York City. At one point on her way through the Bahamas, though, she was repairing a rip in one of her sails and mulling over how to deal with a frustrating, failing rudder when she looked up and saw again a pair of tropicbirds. This time the tropicbirds were "dive bombing," plummeting after a school of little fish near her boat. She thought the tropicbirds sounded so resentful that it was as if they were shrieking to scare her off their waters.

"I only wished they could," Davison wrote, "for this was turning out to be a tedious voyage."

When Davison sailed through this region in the 1950s, some tropicbird populations had been recovering after centuries of European colonization. In distant Bermuda, for example, early laws enacted in the 1880s protected nesting sites and reduced the hunting of eggs for food and the killing of the birds for their plumage, bringing the birds back from local extinction. Today, all three of the global species of tropicbirds are listed as "least concern" by the IUCN, but their numbers overall are declining due in part to habitat loss and introduced mammals, such as rats and cats, that eat their eggs and young.

Davison's boat, meanwhile is back on the water thanks to a

complete restoration by the Northwest School of Wooden Boatbuilding in the state of Washington. In the care of the Community Boat Project, *Felicity Ann* began sailing again in 2018, touring the islands of Puget Sound and sharing Davison's story, inspiring new generations of sailors and adventurers. This time, however, in these northern latitudes, the *Felicity Ann* and its passengers were under the eyes of silent cormorants rather than those of the whistling tropicbirds.

Tuna

Tuna, of which there are currently fifteen species, are among the fastest and most prolific deep ocean schooling fish. Almost unique among other fish, but like mako and similar sharks, tuna have control over their internal body temperature. They keep their bodies warmer than the surrounding ocean water by using a variety of strategies, such as a specially adapted heart and a countercurrent exchange of capillaries similar to that in penguin wings and feet.

The robust silver bodies of tuna are evolved for speed, for constant swimming, and for long-range migrations. The pectoral and dorsal fins of tuna tuck into countersunk slots for streamlining, and their muscles align to taper into a thin caudal peduncle that powers their large crescent-shaped tail fin. With specially located red muscles charged with high-performance, oxygen-rich myoglobin, tuna are ideally suited for deep sea survival and the ability to hunt smaller schooling fish, squids, and patches of crustaceans.

Skipjack tuna (*Katsuwonus pelamis*), with their diagnostic racing stripes on their bellies, are the species today most often hooked and netted around the world for food, especially for canning. Skipjack tuna are harvested on the order of three million metric tonnes a year—and still rising. One of the smaller tunas, skipjacks grow up to 3.6 feet (111 cm) long and can weigh over seventy-six pounds (34.5 kg), but they are normally caught when far smaller. Skipjacks generally range from 40° north to 40° south

Skipjack tuna (*Katsuwonus pelamis*) in the center has dorsal fins and forward ventral fin extended, while the other skipjacks in the school have their fins tucked in for swimming speed.

latitudes throughout all oceans, thriving in a large range of water temperatures from less than 64°F (18°C) to over 86°F (30°C). In addition to their swimming acumen, internal temperature control, and tolerance of different ocean conditions, skipjacks are prolific spawners. One skipjack female can release into the water from 80,000 to 1.25 million eggs *per day*, sometimes for months at a time, which males fertilize with their floating sperm. But skipjacks live fast and die young. About half of the population reach sexual maturity before the age of one, and then most skipjacks only survive on average seven years. This is in comparison to the Atlantic bluefin tuna (*Thunnus thynnus*), for example, which can reach weights of nearly fifteen hundred pounds (680 kg) and live over fifty years.

Tuna only make it to old age if they escape predation by swordfish, marlin, sharks, and toothed dolphins and whales. And their greatest threat, humans, have been killing and eating tuna for a surprisingly long time. Although tuna are deep sea fish, archaeologists believe that forty-two thousand years ago people at the southern edge of Indonesia, from present-day East Timor, pushed off their shores in boats and captured tuna and other open ocean species.

One of the more compelling stories about people and tuna comes from the Maldives Islands, an archipelago of low-lying atolls that barely rise out of the ocean some several hundred miles off the tip of India. Communities on the Maldives have centered their livelihood around tuna for over a thousand years. Recently Dr. Shreya Yadav and her research team out of the University of Hawaiʻi explained that the first human settlers of the Maldives lived in the same way as other island settlers of atolls, where water is scarce and coconut palms and a few other species are the only tenable crops. For protein, perhaps in addition to seabird eggs, the settlers likely first turned to local fish and crustacean species found in the coral reef systems near shore. But then, sometime before 850 CE, visiting Arabian and other international traders found in the Maldives a convenient stopping point when transiting across the Indian Ocean, bound for Sri Lanka or Iran or China or parts of east Africa or anywhere in their global network. The traders discovered that the Maldives were blessed with prolific populations of a small saltwater snail called a cowry, which became, long before the colonial use of wampum in North America, the first truly global, cross-cultural currency. Linnaeus named them *Monetaria moneta*. Like coins, cowry are small, stay shiny, and their available population was

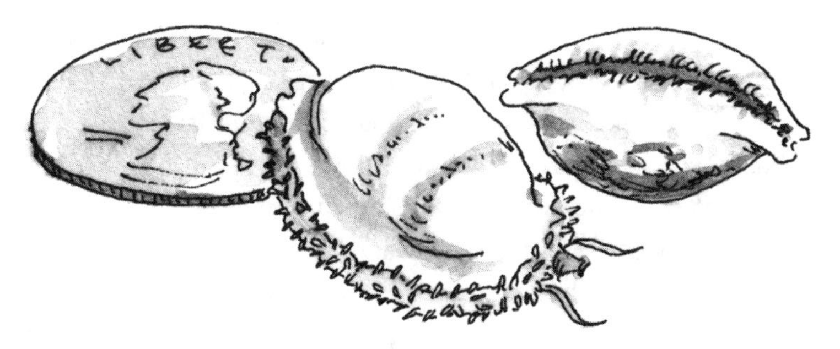

Cowry snails (*Monetaria moneta*) to scale of a US quarter.

finite and controllable throughout given trading centers for a full millennia. Cowry shells became currency throughout the Indian Ocean world, into Africa and the western Pacific Rim, and even into Europe, all the way up into the 1800s. Fishers throughout the Maldives, especially women, raised and harvested the cowries from constructed floats of wood and branches, a very early form of aquaculture.

The cowry trade, because of the regular merchant traffic in and out of the Maldives to get the shells, fit well with the tuna fishery since it provided transport pathways for the fish. Dried tuna may stay edible for months if not years. According to the earliest records of visitors, beginning in 1300s, the people of the Maldives boiled the tuna in saltwater, usually skipjack and less often yellowfin (*Thunnus albacares*), and then smoked large hunks for shipping and storage. In the centuries before refrigeration, these chunks of tuna packaged well, suggesting parallels to the trade in salted cod in the North Atlantic. Dried tuna from the Maldives not only became a staple in local recipes throughout the Indian Ocean network, but the tuna might have also been a regular food source at sea for sailors, supplementing salted meats and sea turtle.

The earliest surviving written record of Maldive tuna is from

1327, left by a traveler from Morocco named Ibn Battuta, who spent four years in the Maldives and wrote, "The food of the natives consists of a fish . . . they call koulb al mâs. Its flesh is red; it has no grease, but its smell resembles that of mutton. When caught at the fishery, each fish is cut up into four pieces, and then slightly cooked: it is then placed in baskets of coco leaves and suspended in the smoke. It is eaten when perfectly dry." Battuta wrote that the tuna was then shipped to India, China, and Yemen. Skipjack tuna is known today in the Dhivehi language of the Maldives as *kalhubilamas*.

The fishers of the Maldives were able to capture tuna so successfully in part because they had a long history of building boats from coconut trees in order to travel from island to island, and due to the oceanography and bathymetry of the region, the Maldivians have never needed to travel tremendous distances to reach deep waters. They have been able to make day trips to capture tuna, for which they fished with poles, lines, and hooks. They bait their hooks with live smaller fish captured in the reefs on the way out, or just chum the water and bring in the tuna on baitless hooks. Some of the fishers of the Maldives still use seabirds, especially noddies, to help them find the tuna schools. Because the Indian Ocean is so affected by stormy monsoon seasons, during which fishing would be more difficult, tuna were prized because they could be dried and kept on hand for months when it is more difficult to go out to sea. So with communities focused on tuna, a likely unintentional environmental benefit in the Maldives was that human communities had less impact on local coral reef ecosystems.

The cowry trade began collapsing in the nineteenth century due to shell inflation (true story). This was around the time Thomas Albro was etching his tuna into a sperm whale tooth in the 1830s and when the naturalist John Davy was returning to London in 1835 to report his news about how these fish have bodies warmer than the water. Meanwhile, tuna continued to be one of the major products in the Maldives for both subsistence and as an export.

Today, though the archipelago is in imminent crisis due to climate change—at an average of only 4.9 feet (1.5 m) above sea level they are the lowest nation on Earth—the Maldives is an independent government, like Kiribati, with an enormous, exclusive economic zone worth millions of dollars exchanged for the privilege to fish, especially for tuna. These days, the right to fish tuna is a currency in itself. The smaller boats still fish primarily by pole and line, eliminating bycatch and ensuring the quality of the fish. So Indian Ocean tuna fisheries by this method have been certified as one of the most sustainable fisheries in the world. But this annual catch of skipjack tuna (and yellowfin to a lesser extent) is a fraction of the global killing of tuna in the Indian, Atlantic, and especially the Pacific Ocean, which is conducted mostly by fishers in large industrial purse seine vessels, who use aggregating devices, diesel engines, hydraulic haulers, helicopters, sonar, and circular nets as wide as sports stadiums.

For now, though, managers believe that skipjack tuna populations, at least, are able to sustain the fishing pressure. But likely not for much longer.

Urchin

(An Interview)

"You're not as cute as a sea otter."

"Neither are you."

"You have to admit, though, you often come up in conversations about sea otters, who really *are* adorable, how the otters wipe their cute little eyes with their adorable little paws—"

"And eat us."

"Right! Isn't it amazing how they grab urchins! If unchecked, you and your friends and family will just munch down a whole kelp forest. Luckily those loveable sea otters dive down and cleverly gather your spiky bodies up and smash you against a rock that they hold in their adorable furry bellies, and—"

"Hey, hey! Enough! This is not starting off as a nice conversation. Nor, Mr. Ocean Bestiary Man, is it an educated one about the intricacies of marine ecology or environmental history."

"You're right. I'm sorry. Tell me about your life as a sea urchin."

"Okay. Well, humans all over the world like to eat our gonads."

"What even are, I mean, ew. Gross."

"Then let's start more basic. The name *urchin* is an old English word for "hedgehog," a little land mammal that is nothing like a pig, but is covered with spiny hairs, similar to a porcupine."

"That makes sense."

"Because sea urchins are round and spiky like a hedgehog?"

"Yes."

"And cuddly?"

"Not really. But the name sea urchin makes some sense now."

"I guess. But sea urchins are not mammals. We're echinoderms. Close cousins to sea stars. Our spines are not like hairs, either. We can move them around, not only for defense but like legs. The spines come out from our internal, oval-shaped shell, the *test*, and we can regenerate our spines if they break. To help walk and to bring food to our mouths, we also have lines of tube feet. You know what else?"

"What?"

"Our mouth is where you'd think our anus would be, on the bottom, while our anus and genitals are on top, where you'd think our mouth would be."

"That's nice for you."

"Yes, it is! There are nearly one thousand different species of sea urchins found in all oceans, from tropical intertidal coastal

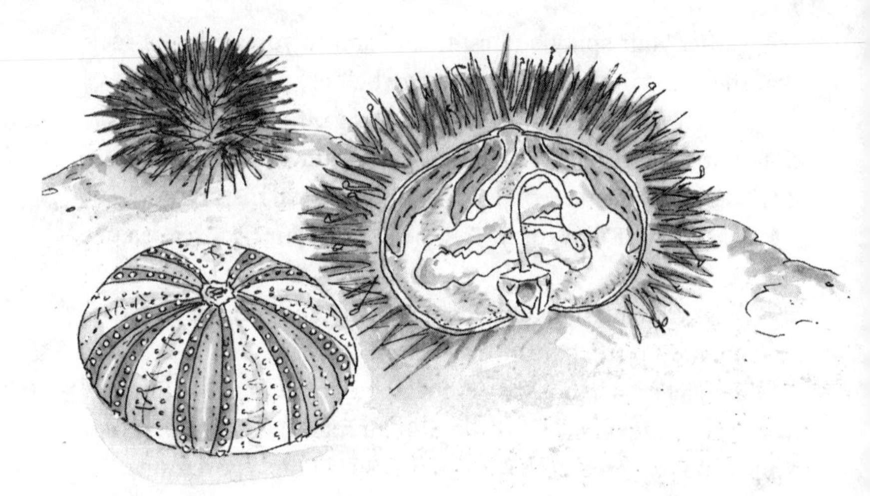

In the foreground, a test, or shell, of a dead green sea urchin (*Strongylocentrotus droebachiensis*) and, right, a cross-section of the internal anatomy. The tongue-shaped gonads are on top (dashed). Aristotle's lantern is the mouth plates at the bottom, and tube feet are among the spines.

waters and coral reefs to the lightless depths of the cold, deep sea. The largest sea urchin species can grow a shell seven inches (17.8 cm) wide, about the size of your average baseball hat, with the spines as thick as pencils. An urchin can live to be a hundred years old! Some sea urchin spines have poison at the end. Most of us are vegetarians, scraping up algae, but a sea urchin will also scavenge animal flesh. Or even each other if things get desperate."

"Yeesh. But a little boring. Can you tell me a story?"

"From Norway, Greece, or Japan?"

"Oldest first, please."

"Or do you want to hear now about my gonads?"

"Oldest story first, please."

"Okay. In about 350 BCE the famous Greek philosopher Aristotle wrote his *History of Animals*."

"Kind of like this ocean bestiary?"

"Sort of. Except a billion times more brilliant."

"Ouch."

"Some historians credit this as the first Western work of zoology."

"That is big. I heard he didn't really understand how paper nautiluses moved."

"Aristotle *did* understand that sea urchins were connected to sea stars. He did careful anatomical work, identifying the five-part radial symmetry of sea urchins. He wrote that the five parts meet at the bottom to form an articulated jaw, looking to him like a small lantern sticking up within the sea urchin body. So today marine biologists refer to our mouth as Aristotle's lantern."

"Cool. But also a bit odd."

"What's odd is that Aristotle passed down folklore that sea urchins swallow rocks to sink down into deep water if a storm is coming. Which is not true. But we do use our suckers and pack tightly into crevices so strongly that waves don't tumble us all over the place. Aristotle, I might add, ate several sea urchin species and explained which parts were good tasting and which were not."

"Don't tell me what he liked best—"

"Gonads."

"Next story."

"Over two thousand years after Aristotle, in the 1750s, Bishop Erik Pontoppidan, wearing a white wig and one of those white ruff collars that looks like a toilet seat, published *The Natural History of Norway*, a bestiary in which he compiled all sorts of common and fantastical animals. His bestiary included mermaids, giant squids that were a mile long, and black and white block prints of sharks and fish that look nothing like real-life animals. A lot like this ocean bestiary, actually, except a billion times more fun to read."

"Double-ouch. Thy spines of speech."

"Pontoppidan wrote that sea urchins are 'one of the strangest animals contained in the sea,' and he recorded all sorts of names for us, like *igelkier, julkier*, sea-apple, and, my favorite, *krake-baller*. He wrote about the beauty of our shells, the beautiful green luster of our spines, and he also did dissections in which, when describing our insides, he used the word *slime* a lot."

"Sounds reasonable."

"Although Bishop Pontoppidan passed on the myth about sea urchins swallowing stones before storms, he did write one of my favorite sentences ever, as translated from the Danish: 'The Sea-urchin is found on a sandy bottom, and rolls himself about with his prickles, wherever he pleases.'"

"That *is* lovely."

"Pure Pontoppidan poetry."

"I hate to ask this, but did he eat sea urchins, too?"

"He did not personally, which I appreciate, but Bishop Pontoppidan did report this secondhand: 'Sicilians, whose taste must be different from ours, reckon this creature to be delicate food; they break the shell, and eat the inside raw with spoons.'"

"Were the Italian people eating—"

"Yes, gonads! Be patient now, though. Let me tell my last story. Since perhaps as early as the time of Aristotle, people on the other side of the world, on the coasts of Japan, were also harvesting sea urchins, including women known as *ama*, divers with a similar style to that of the haenyeo in Jeju, Korea. Although sea urchins are, ahem, fairly easy to capture, there is not a lot of edible food inside our shells. So urchins have always been a delicacy in small quantities. In the 800s it was on an official list of offerings to the Imperial Court. In the Edo Period of the Tokugawa Shogunate (1603–1868), urchin was one of the 'three great rare tastes under heaven.' The other two were sea cucumber guts (another echinoderm, not to be confused with a pyrosome) and mullet roe. Urchin was a food reserved for royalty and special occasions, and it does not keep well. It was not until the 1950s in Japan and then with the popularization of sushi around the world, that Japanese style urchin found a much larger foodie audience."

"These are gonads, I take it? I'm ready now."

"The food of the urchin in Japan is called *uni*. When humans eat urchin, usually raw, from nearly any culture, it is this uni. Some even still scoop it straight out of the urchin with a spoon. Uni is usually bright orange, soft, and squishy. Most humans like to imagine it is roe, fish eggs, like caviar, but in truth uni is the gonad of the sea urchin, an organ that includes either the unfertilized eggs or the undeveloped sperm of the urchin, the ovaries *or* the testes of the animal."

"Does it, um, taste good?"

"How do your internal organs taste?"

"Touché, urchin."

"So now you know. That wasn't so bad, was it, Mr. Ocean Bestiary Man?"

"Have a nice day, Mr. Krake-baller. Hope those ugly, mean old sea otters don't get too many of you."

Velella and the Man-of-War

In the summer of 1879, there was no Panama Canal, so Morton MacMichael III, traveling aboard the merchant ship *Pactolus* from Philadelphia, was settling into a voyage of four months or so around Cape Horn and all the way back up north to California. He later polished and published his journal as *A Landlubber's Log, of his voyage around Cape Horn* (1883).

About three weeks out, they were sailing in the tropical Atlantic Ocean before crossing the equator. MacMichael had seen (and eaten) his first flying fish, and he'd enjoyed his first glimpse

Ship's cat (*Felis catus*) and a Portuguese man-of-war (*Physalia physalis*).

of the Southern Cross. One morning the ship steered through a collection of strange animals on the surface.

"Passed through a large fleet of nautilus," he wrote, "those renowned little creatures of the jelly-fish species, that spread their tiny film-like sails in delicate shades of pink and blue, and cruise about over the waves, sometimes alone or in little groups, and again, as I first saw them, in vast numbers. The sunlight playing on the thousands of rising and falling sails made a very pretty picture."

Since he was a passenger and didn't have any work to do, he leaned over the side with a bucket on a rope and tried to scoop up a specimen. After a while he caught one of the delicate animals and brought it on deck. When he turned away from his catch to fetch the carpenter to show off what he caught, one of the ship's two cats darted over, dipped in her paw, and ran away with it. The cat dashed across the deck then dropped its prize with a yowl, running into the galley "as though a dozen dogs were at her heels."

MacMichael wrote, "During the rest of the day she sat in a corner, uttering plaintive meows, and alternately rubbing her cheeks on the deck or scraping her swollen tongue with one of her front paws."

Despite the confusion of common names, MacMichael's animal in the bucket was not one of the paper nautiluses, those female octopuses riding in their shells. He had instead scooped up a Portuguese man-of-war (*Physalia physalis*), which is one of two types of sea jellies, free-floating, gelatinous hydroids, that genuinely sail their way across the surface of the water. The second kind is smaller and lesser-known to those ashore, and, at least by this description, were likely also to appear in this fleet that MacMichael sailed through. This second sea jelly is called a by-the-wind sailor, but also just called velella, after the scientific name, *Velella*, which means "little sail." Both of these sea jellies—the Portuguese man-of-war and the little velellas—are found floating on the surface in warm waters, but they can live in colder latitudes, too, usually when blown by strong winds and carried by the currents.

Sea jellies are 95 percent water and mostly clear. Both the velella and the man-of-war are actually colonies of different individuals that act as one, separate living parts of the larger animal with specific jobs, similar in this way to the pyrosome.

The man-of-war, as MacMichael described, can be tinted pink, blue, and purple. Some people call them "blue bottles." Their floating bodies can grow up to one foot (.3 m) long, with tentacles that can spread down into the water over fifty feet. On its tentacles are stinging cells, or nematocysts, which capture and

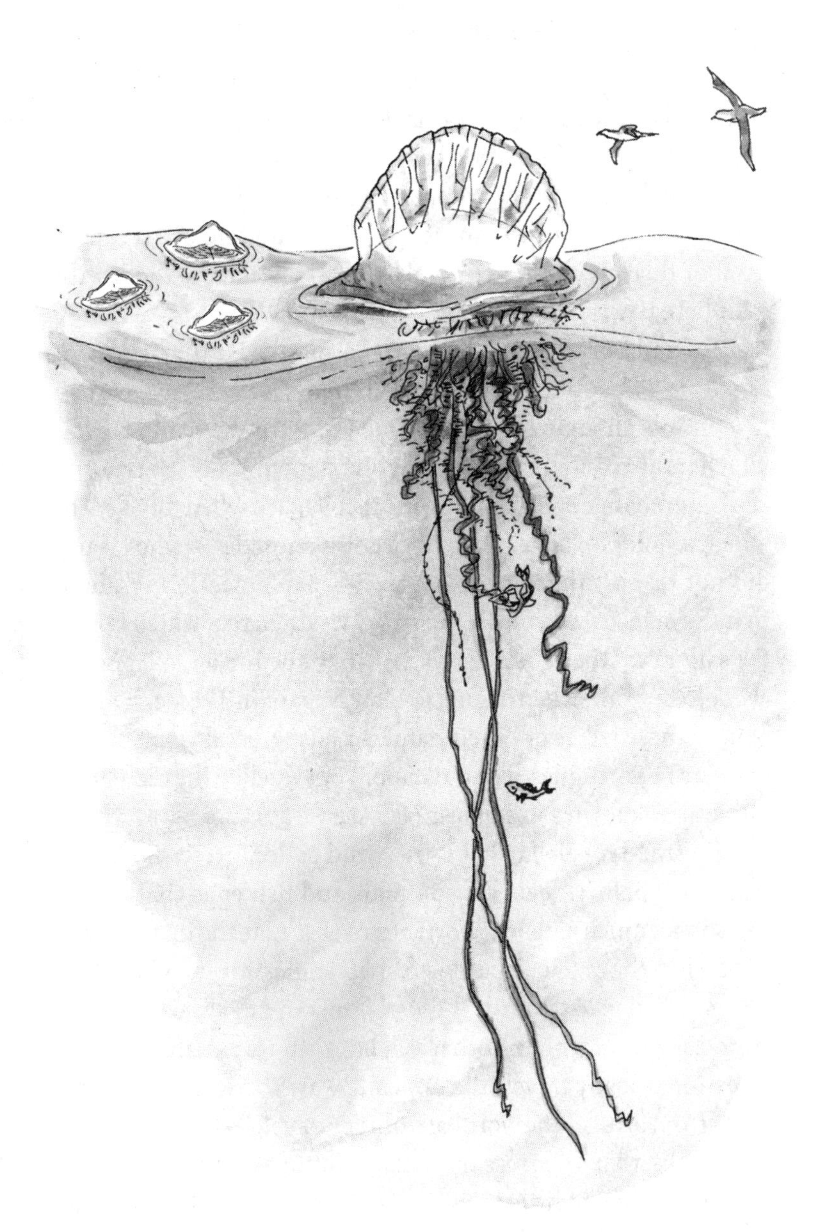

By-the-wind sailors (*Velella velella*), left, and a man-of-war (*Physalia physalis*).

kill their food, mostly small fish. These stingers can kill a human adult, even if the man-of-war is dead. You might have suffered a few itchy stings from their tentacles if you've been swimming at the beach at an unlucky moment. Though velellas do not sting nearly as much as the man-of-war, if you do find them on the sand or floating in the water, it's best to not handle them with your bare paws.

The "sail," or upper body, of velella and the Portuguese man-of-war is a float filled with gas. Another traveler at sea, named Joseph Ingraham, wrote in the 1830s of a trip he made during which he also caught a sea jelly (with no mention of a feline victim). Ingraham noticed how the float collapsed when touched, to which a sailor on board looking over his shoulder—who clearly did not respect the Portuguese navy—explained to Ingraham that the man-of-war deflates in heavy weather, too, which is how the sailor said they got their name: "They takes in all sail, or goes *chuck* to bottom, when it 'gins to blow a spankin' breeze." Others believe these jellies got their name because their floats look like helmets of Portuguese conquistadors or even like the old Portuguese ship known as a *caravela redonda*.

The smaller velella, or by-the-wind sailor, grows no larger than your palm. It eats tiny animals and fish eggs that it captures in its much smaller, shorter tentacles. Like the man-of-war, the sail of the velella is at one of two angles for its whole life, like a little paper sailboat. It mostly sails in one direction away from the wind, while another velella, with its sail in the other direction, mostly travels the opposite way. There are regular reports from all over the world of millions of velellas washed up on beaches, perhaps because of a change in the wind direction and currents: a fleet of little velellas that cannot steer off the rocks.

Scientists do think, however, that both the velella and the man-of-war have some control in how they adjust themselves to the weather. Velella in particular have evolved an aerodynamic structure to their sail-like bodies. In order to survive in heavy winds, they are triangular (so the pressure is closer to the surface), slightly flexible (consider how a tree bends in a storm), and its base is shaped in an "s" to increase its strength (try this with a piece of paper).

MacMichael wrote that after the sea jelly incident, the cats aboard the *Pactolus* returned to their wailing and meowing along with the sailors playing music on deck. I wish they had known a more recent song from the 1960s by Flanders and Swann, which certainly would have been the cats' favorite tune. It goes like this:

> I do not care to share the seas
> With jellyfishes such as these,
> Particularly . . . Portuguese.

Walrus

The walrus (*Odobenus rosmarus*) dives comfortably under the frigid surface of the Arctic Ocean because the animal is protected by a thin layer of light-brown or blond hair over leathery, wrinkly skin that can be 1.5" (3.8 cm) thick. Underneath their skin, walruses have an insulating layer of blubber that can be over 4" (10.2 cm) deep. And the old bull males, much larger than the females, can weigh over 3,300 pounds (1,500 kg). An enormous rotund body, as it does for the right whale, reduces the surface area that would cool the animal in the water.

Hunting walruses one summer day in 1908 was an Arctic explorer named Matthew Henson, leading a group of shipmates. They were in Ikeq-Smith Sound, an extension of an ice-ridged, open swath of sea between the far northern tips of Nunavut and Greenland. Henson knew that the walruses surfacing and diving here were not primarily hunting for fish. Usually diving at most two hundred and fifty feet or so (76 m), walruses feed almost entirely on shellfish—cold water clams, mussels, scallops—and crustaceans, such as shrimps and crabs. Henson knew their diet because he was a reader. He had also spent a lot of time butchering, opening their stomachs, and learning from the Indigenous people of the region.

Henson might not have known, however, before the age of underwater photography and dry suits, that walruses do not use their tusks for finding food. Close relatives to seals and sea lions, walruses use their flippers to clear the bottom as they burrow their thick whiskers into the silt to suction out the soft bellies of mollusks. Both male and female walruses have tusks, which are two enormous canine teeth that grow throughout their life, growing in the males to be longer than three feet (1 m), longer than most people's arms, even with the fingers outstretched. Walruses use these tusks to compete with each other, for defense against predators, such as polar bears and killer whales, and for helping them get up onto the ice floes, using their tusks like grappling hooks.

Already above 75° north latitude, Henson and his crew made their way farther north with the steam schooner *Roosevelt*, pausing here at the mouth of Ikeq-Smith Sound to do some hunting. Matthew Henson secured his gun. He looked around. He also occasionally gazed back up to the ship, to see if his shipmate up in the ship's crow's nest has spotted any walruses. Henson, an African American man, was orphaned as a child. He started his career at sea at twelve years old as a cabin boy on a ship bound from Baltimore to China. By this point, Henson was Commander Robert Peary's right-hand man and a veteran of several Arctic expedi-

Two walruses (*Odobenus rosmarus*).

tions. Henson was a gifted carpenter, sailor, hunter, and logistics manager. He was fluent in Inuktitut, the Inuit language, and he was respected by both the Inuit and the American crew as the expedition's most skilled sled driver and dog handler.

Henson and a couple of American shipmates, along with a few Inuit men, hunted walrus for meat to fortify their dogs because they planned to winter over at the absolute northernmost spit of land, Cape Sheridan. At first Henson and his crew used a motorboat that day to hunt walruses, but the engine sputtered out, so they rowed. Henson shot and killed two walruses. Blood clouded the clear, icy waters. The men shed layers of clothes as they towed the walruses back to the ship. After hoisting the carcasses ashore and into the hold, Henson led them back out, this

Atlantic walruses burrowing into the bottom for the Arctic truncate soft-shell clams (*Mya truncata*), or *ammuumajuq* in Inuktitut.

time with a whaleboat and more oarsmen. They killed four more walruses.

Over the following days, as they continued to hunt and allocate the meat in different ways, using every single part of the animal in some fashion, Henson took particular care with two of the walrus skins, which he had been asked to prepare to send home to be stuffed, presumably for museum taxidermy.

Henson and the others spent a long winter, braving temperatures of 50°F below zero and constant darkness at their camp. Their dogs survived on the walrus meat they had caught on the way north. When the weather eased, the team headed north on sledges pulled by their dogs, a perilous journey across uneven ice and occasional leads of soft slushy ice. One man fell through the ice and drowned. Eventually, Peary, Henson, and four men of the Inughuit tribe of Inuit, named Ootah, Eningwah, Seegloo, and Ooqueah, battled at last to the spot Peary calculated to be the North Pole, arguably becoming the first humans to do so. Henson and the others wore clothing of polar bear fur and sealskin that was sewn by Inuit women.

In his story of the expedition, *A Negro Explorer at the North Pole* (1912), Henson left a series of small, intriguing observations about walruses, especially in relation to the dwindling tribe of Inughuit people living at the far northwest of Greenland. American and European ships coming up to the North Water in the 1800s and early 1900s had been hunting bowhead whale, a few species of seal, and walruses in great numbers. Although he and his shipmates killed at least seventy walruses themselves during their expedition, and several of Henson's observations about the Inuit people are wincingly condescending, Henson reflected on this visit to the region: "It is sad to think of the fate of my friends who live in what was once a land of plenty, but which is, through the greed of the commercial hunter, becoming a land of frigid desolation. The seals are practically gone, and the walrus are being quickly exterminated."

Henson is far from the first visiting mariner to record colonial devastation on marine populations and its impact on Indigenous peoples. But Henson's comments are exceptional because the more widely read parallel accounts of the same American polar expeditions written by Commander Peary and Peary's wife, Josephine Diebitsch Peary, for example, include almost no comments about marine mammal depletion or its impact on the local people. These observers instead celebrated how their style of killing with firearms and harpoons from whaleboats was far more effective than the Indigenous methods conducted from the edge of the ice. Yet Henson's experience as an African American from the segregated United States, as explained by scholar David Anderson, might have allowed him to see beyond national or colonial pride in his expeditions. Working more closely alongside people who were also marginalized, mariners of color like Henson, especially those who came in regular contact with Native communities and their ways of life, might have been able to more keenly empathize and better perceive the unfolding of environmental degradation. There's a reason that Inuit called Henson, in translation, "Matthew the Kind One."

The hunting of walrus had been a significant and central part of Indigenous life around the North Water for over four thousand years. Walruses, *aaveq* in Inuktitut, are the largest of the Arctic pinnipeds, and their numbers and migrations have been historically more predictable than those of bowhead whales. Different groups of humans had moved in and out of the North Water region for centuries, with different traditions and hunting practices, and Western archaeologists and oral traditions show that walruses have always been part of Indigenous survival in the Arctic for over two thousand years: for meat, for tools from the tusks, for rope from the skin, and for oil for light from the blubber. Walruses have also been central figures in Indigenous storytelling and the Inuit people's spiritual lives in this part of the Arctic—reflected in their traditional clothing, musical instruments, artwork, and cosmologies.

The early Indigenous tribes and the later Inuit cultures might have impacted walrus populations to some extent, but certainly not in the way that European hunting did. Beginning about 980 CE, the arrival of Norse hunters to the coasts of Greenland utterly shocked the walruses. Within a couple of centuries, these Viking settlers decimated local walrus populations. Then they and later European people hunted walruses farther west, throughout the Canadian Maritimes. With the additional use of firearms, colonial hunters recorded and slaughtered herds of walruses by the thousands in the icy regions of the Gulf of St. Lawrence and as far south as Sable Island off Nova Scotia. This commercial hunting continued into the early twentieth century. By Henson's day, and still over a century later, walrus live only as far south as the Labrador Sea and Hudson Bay.

Matthew Henson died in 1955 after finally getting some of the honors and respect that he deserved for his expeditionary accomplishments. If Henson visited the North Water region today, he would be astounded by the reduction of the ice due to man-made climate change. Henson's coal-belching steamship *Roosevelt* would mean something quite different if it were now grumbling up toward Ikeq-Smith Sound. This loss of ice has affected the Inuit people in several dramatic ways, including their relationship with one of their primary resources, the walrus, since these animals regularly haul out on floating ice pans and on ice edges, especially the females and their young, who find safety on these cakes of ice.

At the start of the 2020s, walrus populations seem relatively stable in this part of the Arctic, but they remain only a fraction of their historical numbers and range. The Inuit elders speak about past walrus abundance: when there used to be collective hunts with rope and harpoons, the men stretching out, single-file on the thin ice toward the thinner edges to reach ice holes where walruses poked out their heads to breathe. There are now roughly thirty thousand individuals of the Atlantic subspecies of walrus. In the Arctic of the North Pacific, there are over four times as many walruses of the Pacific subspecies. Increased open water throughout the Arctic, however, means more shipping, fishing, mining, and tourism, which means more disturbance to walruses, whose females only give birth to one calf every three years. A crack of hope, though, is that the management of walrus hunting in recent years has been planned with more input from Inuit representatives. And today only the Indigenous peoples are allowed to hunt walruses in the North Water region and with strict quotas that they help set.

Wandering Albatross

This was her first voyage at sea. She had just turned twenty-three years old, and now she was leaving home to sail with her new husband, the captain of the whaleship *Tiger*. As they sailed down into the far South Atlantic, toward Cape Horn, Mary Brewster still felt seasick sometimes, but she had started to grow accustomed to her life on the ocean. She had learned by then how to stand athwartships with her feet planted widely, and how to lean when the ship rose, heeled, then rolled off a big wave from astern.

Brewster wrote in her journal nearly every day. This account of her voyage is the earliest complete record by an American woman aboard a whaleship. It's also a particularly useful manuscript for learning about our relationship with one of the most iconic of seabirds.

On Monday, January 19, 1846, a few days north of the Falkland Islands, the air was warm, the sea unexpectedly calm.

"This morning one of the sailors caught an Albatross," Brewster wrote, "which gave me a good chance of seeing one. I was quite surprised in its size, being much larger than it seems when in the water."

For the previous few days, the sailors had been trying to catch seabirds with a piece of pork on a fish hook, and now they had snagged this albatross. The men dragged the seabird on deck, still alive.

Brewster wrote the following description of the bird: "The

Juvenile wandering albatross (*Diomedea* sp.) struggling on a ship's deck.

most peculiar parts were its wings and bill, the former very long and narrow, the latter very strong and hard, straight to near the extremity when it suddenly curves. The upper part appears composed of many pieces furrowed on the sides and crooked at the point. The lower smooth and short. The feet short, the three toes long and the feathers Brown speckled with gray. The bill is of a pale yellow. After being on board a short time it had every appearance of seasickness, so they let it go, after which it was as lively as ever."

Brewster didn't write down exactly the size of the albatross, so it's difficult to identify exactly what species it was. The brownish-gray speckles suggest it was almost certainly a juvenile, and, since they were on the edge of the Southern Ocean, it was likely one of the wandering or royal albatrosses (*Diomedea* spp.), all of which have yellowish-pinkish bills and can have wingspans between eight and nearly twelve feet long (2.4–3.6 m). Their bills do indeed look like they're composed of multiple layers, and those furrows on each side that Brewster described are channels by which salts drip down the beak after being excreted out the tube-like nostrils. Albatrosses are part of a larger grouping of seabirds known as tubenoses, the adaptation that these animals have evolved that allows them to stay offshore for so many months at a time, part of a system that separates and discards the excess salts so the birds can process seawater in their bodies.

As early as 1594, the English privateer Sir Richard Hawkins and his crew were catching albatrosses with a fishing line in the open Southern Ocean. Throughout Brewster's 1800s it was common practice for Western sailors to capture albatrosses and other seabirds by the method used on the *Tiger*: streaming a piece of line astern with a baited hook or a sharpened piece of metal. Sailors did this into the early 1900s. Aboard some ships, passengers, crew, and naturalists shot at seabirds with guns and might even launch small boats to go retrieve them if the conditions were safe.

What's curious is that Mary Brewster did *not* mention in her journal the famous English poem, "The Rime of the Ancient Mariner" (1798), which is about all the things that happened to a man after killing an albatross at sea. The poet, Samuel Taylor Coleridge, spun the fantastical ballad in part from a true story that he read about a sailor killing a "black albatross" off Cape Horn, not far from where Brewster was sailing that day. In his poem, a mariner shoots the albatross with a bow and arrow, because, it seems, he thinks it is the cause of their bad weather. Conditions get still worse, though, so his shipmates blame him for killing the bird, and, before they all turn into what we would now call zombies, make the mariner wear the enormous bird around his neck. All sorts of horrific things continue to happen, until the sea snakes come in to save the day. In short, the ballad teaches that it is an exceptionally bad idea to shoot an albatross. For over two centuries, this poem was read and even memorized by hundreds of thousands of schoolchildren and published in poetry collections for over two centuries in English-speaking countries around the world. Yet the popularity of the ballad didn't seem to affect sailors' practices all that much. Brewster never mentions the poem, and, though a surprising number of sailors were well read, it seems it did not influence too many voyagers. Several well-read sailor-writers who went to sea from the 1840s to the 1930s, such as Richard Henry Dana Jr., Francis Allyn Olmsted, Anton Otto Fischer, and Eric Newby, all describe shipmates catching albatrosses, yet with only a passing comment or no mention at all of "The Rime of the Ancient Mariner."

Sailors and passengers back then often liked to catch albatrosses simply because the birds were so enormous and unique. Brewster, too, wanted the experience of seeing one up close. Sailors were interested in the long feathers or the beak as souvenirs, and some mariners with an artistic bent crafted walking-stick handles from the beak, sewed tobacco pouches with the birds' webbed feet, and fashioned pipe stems out of the hollow bones. Pacific Island peoples, such as those in Hawai'i and in Aotearoa New Zealand, have used albatross feathers for adornment, ceremonies, ornamental staffs, and decorations for their boats; the bones were used for flutes, as tools for tattooing, and for other crafts.

Today, the majority of the twenty or so albatross species that live around the world's oceans are vulnerable or endangered, although not because of sailors. Albatrosses are trying to survive the loss of many of their isolated island rookeries due to human settlement, sea level rise, and introduced mammals, such as rodents and cats. Albatross populations also suffer due to ocean plastics, and, although there has been genuine recent improvement in gear and practices, these birds still continue to drown by the thousands each year after hooking themselves in the longlines of deep sea fishing boats, much in the same way that albatrosses were attracted to the lures streamed astern of the *Tiger*.

Three years after her first close encounter with an albatross, Mary Brewster and the whaleship *Tiger* were now homeward bound in the Southern Ocean after whaling throughout the Pacific. Sailing toward Cape Horn, this time from the west, Brewster stood on deck in freezing rain, hail, gales, and headwinds, and each day she recorded schools of pilot whales, dolphins, and flocks of birds.

On Thursday, December 2, 1848, Brewster wrote, "Caught an Albatros tied a bottle to his neck with a note informing anyone who takes him of our welfare; after securing the bottle let him go. when last seen was steering straight for Chili."

It turns out there used to be a small tradition of this, too, of sailors tying messages to albatrosses. For example, in 1842 whaler John F. Martin wrote that they caught an albatross and put a leather label around its neck, with the name of their ship and their position. In 1857 a whaler's wife named Henrietta Deblois wrote of an albatross she observed from afar that had a blue ribbon tied around its neck. Another time a whaleship captain named Hiram Luther shot an albatross quite near where the *Tiger* had labeled theirs. When Luther brought the bird on board, he found that it had a vial around its neck with a note from another whaler. The other captain seemed to have been bored, complaining in the message that he hadn't seen a whale in four months. Based on the position and the date of the note, that albatross had flown about thirty-six hundred miles in twelve days. In 1888 a captain reported a note-toting albatross captured on the coast of Western Australia. This albatross's message was from the crew of a French vessel shipwrecked on the Crozet Islands, which was

Adult wandering albatross (*Diomedea* sp.) in the Southern Ocean off Cape Horn.

almost on the other side of the Indian Ocean. And as late as 1906, shipwrecked men on those same Crozet Islands decided to try to get help using albatrosses, sealing small messages into metal canisters and tying them to the legs of sixty birds. (It didn't work, but luckily they managed to flag down a passing steamship.)

Brewster returned from her first voyage aboard the *Tiger* and turned right around for a second one. Outward bound in the middle of the Indian Ocean, she wrote in her journal, "The sailors are amusing themselves with catching Goneys which they intend cooking in the try pots, tough mess but anything new goes here."

The crew were eager for anything besides their normal "salt horse" and ship's biscuit. And this nickname of goneys or gooneys for albatrosses is still around. It might have come from the bird's awkward, seasick-like nature on deck (think goofy, a "goon"). Sometimes sailors could be quite cruel to the birds when they were on their ship, even if they planned to let them go.

Yet Brewster had discovered on her voyage that albatrosses are anything but goofy in the air. Their flight is jaw-droppingly skillful, spellbinding really, in their ability to glide effortlessly on long, thin wings with barely a flap in the harshest of wind conditions. They have a special adaptation to lock their wings in place when soaring. On his bestiary tooth, Thomas L. Albro drew these characteristically extended, flat wings for his albatrosses soaring over the French Rocks. Juvenile wandering or royal albatrosses, like the one Brewster first described so carefully, will spend up to five years at sea without coming back ashore, flying across tens of thousands of miles of open water before returning to the island of their birth to mate. With their long, light wings angled at times perpendicular to the seas, they search for carrion or prey with a trait rare among birds on land: a strong sense of smell that allows them to locate food from miles away.

After Mary Brewster's two voyages on the whaleship *Tiger*, totaling over five years away from home, historians are not certain if she went to sea again, although her husband did go back and forth around Cape Horn a few more times as a captain of merchant ships. Brewster died in 1878 in her hometown of Stonington, Connecticut.

What would she have made of the comments of Robert Cushman Murphy, a naturalist going to sea for the first time himself in 1912 on one of the last New England whaleships under sail. Murphy wrote in the diary he kept for his fiancée, Grace Barstow, "I now belong to a higher cult of mortals, for I have seen the albatross."

Whale Shark

Once there was a wealthy man named Charles T. Brooks from Cleveland, Ohio, who traveled to Florida to go fishing. It was 1912, and he chartered a sloop named the *Samoa*, run by Captain Charles Thompson, a man who sometimes wore a white skipper's cap and was famous in the area for catching sharks and any other kind of big game in the ocean. The two men and Captain Thompson's first mate, Bob Denny, sailed south from Miami along the Florida Keys. Brooks was hoping to fish for some big tarpon, otherwise known as silver king, in similar waters as Harry Dean's big catch, Captain Bush's green turtle escape, Hemingway's sharks, and Audubon's noddy.

The *Samoa* sailed south and anchored off Knight's Key. In the morning the three men spotted the largest fish anyone in the Florida Keys had ever seen. According to the locals, it had been swimming around the area for a few days. But no one knew exactly what it was.

Captain Thompson thought it was a shark. He vowed to catch it.

They approached in a small boat, rowing directly over the massive, spotted fish, until they plunged a harpoon into its back. As the morning wore on, they speared the animal several more times, shot it in the back about fifty times, and lashed it all around with rope. Crowds of people gathered to watch, overlooking from the viaduct.

"I was surprised that the fish did not put up any fight," wrote

Brooks in a personal letter afterward. "He proved to be a sluggish monster, and seemed to fail to realize that anything particular was happening to him."

Though the fish did not thrash or fight or even seem to be in pain, it took the three men, with the help of several other eager helpers, about eight hours to get the poor animal onto the beach. When it finally died on the sand, they measured it at thirty-eight feet long and eighteen feet around 11.5 × 5.5 m). Imagine about seven full-sized refrigerators long by three refrigerators wide.

Captain Thompson and several other men lashed the fish to the *Samoa*'s port side. Powered by a tugboat on their stern, they towed the animal back up to the mainland. Once in Miami, they floated their catch onto a marine railway. But the weight of the fish broke the rails. It likely weighed over thirteen tons (11.8 mt).

The *Miami Metropolis* and other newspapers spread the story, although still no one knew what this gigantic fish was exactly. The marine biologists at the Dry Tortugas, the outermost island of the Florida Keys, thought from the newspaper description that it must have been a huge killer whale.

Although they took photographs of the catch—including one picture of the fish lashed beside the *Samoa* with a grinning man literally sitting inside the animal's mouth and another photograph with two men wearing button-down shirts and derby caps sitting on top of the fish's head—these images did not yet make it into the newspapers or to the biologists who might be able to identify what this enormous thing actually was.

At this point Charles T. Brooks went home to Ohio with quite a sea story. Captain Thompson showed the fish in Miami for a few days and then poured gallons of salt water and formalin on the carcass to try to preserve it until he was able to hire a man to help him properly taxidermy the animal under a purpose-built shed.

Meanwhile, a biologist named Eugene W. Gudger, a professor from North Carolina State Normal College was a visiting researcher at the Dry Tortugas. He read the newspaper account and was eager to identify this animal. Gudger wrote to Captain Thompson and visited the carcass in Miami. Gudger confirmed that because of its skin texture, fins, and the distinctive gill slits: this was a type of shark. More specifically, he declared this to be a whale shark (*Rhincodon typus*), which is not only the largest of the sharks, it is the largest fish in the entire ocean and the largest vertebrate on Earth today that is not a mammal. Whale sharks have enormous gills, and a huge, flat mouth that looks a bit like a puppet's. On their dorsal side whale sharks have a distinctive white or yellow polka-dotted pattern with thin pale lines. The pattern is so regular that in Mexico whale sharks are known as *dominós*, after the tile game.

The short visit to the whale shark's carcass was not enough for Gudger. He wrote several more letters to Thompson and Brooks to get details. But he was never able to see the full stuffed fish in Miami—or ever, as it would turn out.

Captain Thompson's catch resulted in one of first published accounts of a whale shark, which had been first described in modern Western records only ninety years earlier by a medical doctor in South Africa. Even after Thompson's whale shark and a few highly publicized encounters written by William Beebe and Thor Heyerdahl in the 1930s and 1950s, marine biologists still knew very little about this fish in the wild until the 1990s, when, for example, scuba diving biologists such as Eugenie Clark, when she was not examining electric rays, was able to observe them during seasonal feeding events.

Whale sharks are solitary, deep-sea creatures that live in tropical waters in all oceans. They have been observed as far north as Japan and Nova Scotia, and as far south as the Cape of Good Hope and off Sydney, Australia. Gudger went on to spend a surprising portion of his life mapping and collecting reliable whale shark sightings. (Gudger lived a life of surprising and eccentric ichthyological enthusiasms, as you'll see in the following story about *Xiphias gladius*.)

Whale sharks have as many as three hundred rows of tiny teeth in each jaw, each tooth smaller than the size of pencil point. The three thousand or so teeth are usually under a skin flap. It's not clear how the sharks use these teeth, if they're vestigial, because whale sharks primarily feed by opening their mouth to suction up swaths of ocean filled with krill, copepods, and small fish, which they sieve through twenty spongy, porous pads in their throat while forcing the water out their gills.

Among the five hundred or so species of sharks, only three of them are plankton feeders like the whale shark. These include the basking shark (*Cetorhinus maximus*), which can grow almost as large, and the megamouth shark (*Megachasma pelagios*), which only grows to some eighteen feet (5.5 m) long. Whalers in the 1800s, such as Thomas Albro, often referred to this group of sharks as "bone sharks," because the filtering plates in the sharks' mouths reminded them of whale baleen, which the whalers, clearly to confuse future generations, also called "bone."

Captain Thompson's specimen was not the largest whale shark on record. While diving off Baja, California, Eugenie Clark observed a whale shark that was fifty feet (15.2 m) long. There have been unsubstantiated stories that claim even longer whale sharks, perhaps up to seventy feet (21 m). Recently, scientists have tracked whale sharks diving over a mile beneath the surface and have found individuals living to be some seventy years old.

The kookiest part of this whole whale shark story did not end in Florida.

Gudger could not find Captain Thompson or the whale shark when he planned to meet them in Miami, because after Thomp-

Whale shark (*Rhincodon typus*).

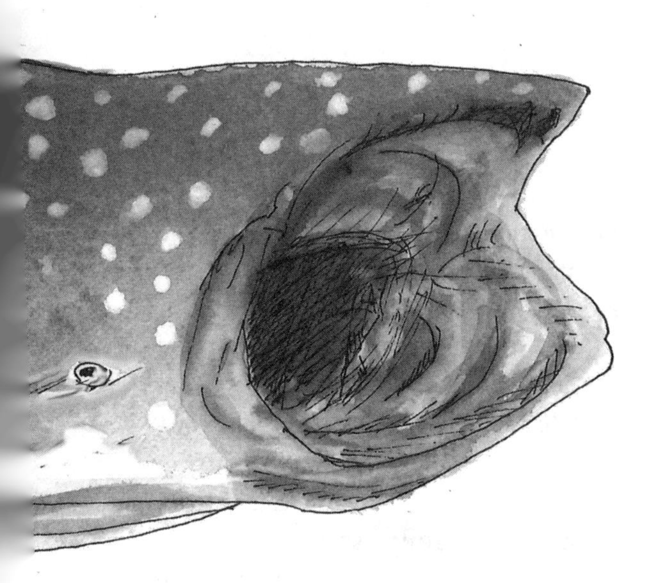

son and his taxidermist finished preparing the specimen, the captain took the fish on the road. On the rail, actually, to be more specific. He exhibited his prize catch on a railroad cart all over the country—from Florida to Atlantic City (where Gudger hoped again to catch up with him, but was too late), and then on to Chicago, charging admission wherever he went. Photographs from his whale shark on a rail car exhibit still exist as postcards. In one, two small children, half as tall as the dorsal fin, straddle the top of the whale shark's back while a man with a bowler and a black mustache, presumably Captain Thompson himself, stands toward the tail looking at the camera.

The Midwest whale shark spectator market satiated, Captain Thompson then shifted the carcass onto a boat named the *Tamiami* to exhibit the specimen to people along American rivers. The *Tamiami* and the whale shark were eventually lost to a fire in 1922. However it happened, it was likely an act of mercy, because the carcass was by then worn and ragged and rumored to smell rancid.

Today, out in the open ocean, biologists believe the whale shark global population has been reduced by half of its population in only the last seventy-five years, due mostly to people killing them for food or to their death after accidental entanglements in fishing nets or impact by ships and propellers. In a few places in the world, such as off the Yucatán Peninsula in Mexico, however, there is conservation-minded ecotourism in which you can swim with whale sharks while helping to fund research and local communities. Dr. Gudger would be thrilled to know, too, that now there is a global database called "Wildbook for Whale Sharks," to which you can send photos to help identify and report individuals. More and more countries are establishing protected areas out at sea and regulating fishing methods in their waters. For example, in 2021 officials in Mozambique in East Africa granted legal protection to whale sharks and other large fish.

Xiphias gladius

"It appears to parade up and down in front of the window, dashing to and fro, not knowing how to interpret our appearance, swimming a few meters and then returning, as though fascinated by our great Plexiglas eye. Suddenly it attacks."

So writes the Swiss underwater explorer Jacques Piccard. It's the summer of 1969, and he and four other men from the United States and the United Kingdom are about 825 feet (250 m) below the surface, floating northbound in the Gulf Stream in a 48-foot (14.5 m) long submarine named the *Benjamin Franklin*. Piccard is tall and slim with dark, short hair and a thin beak of a nose. He and the four other men are peering out into the sea, which was lit by a floodlight mounted outside the hull. They are watching their submarine be attacked by *Xiphias gladius*, the broadbill swordfish.

The swordfish seems to be aiming at the window, but the animal—deep blue with a tall dorsal fin and an enormous eye—spears its bill against the steel, just underneath the plexiglass, glancing off. The men hear the impact, a clang, from inside.

"Why do these swordfish attack submarines?" Piccard writes later. "Are they fascinated, hypnotized by the portholes? If they mistake submarines for monsters, which perhaps they actually are, what courage these fishes show in attempting to impale an adversary so much bigger than themselves."

Out their window, Piccard and his crew see a second swordfish in the distance, but the animals do not attack the submarine again. Piccard explains that only two years earlier, a much smaller submersible named *Alvin*, operated out of the Woods Hole Oceanographic Institution, had also been speared. In their case, at a depth of about two thousand feet (610 m), the swordfish rammed itself so hard into a fiberglass joint that it stuck itself all the way up to the base of its sword. This eight-foot-long (2.4 m) *Xiphias gladius* had to be brought back up to the surface, still alive. The swordfish could not be freed without breaking off its entire bill. The scientists euthanized the fish—then ate it.

Xiphias gladius are some of the most global fishes on Earth, swimming in all oceans from as far north as Newfoundland and Japan to as far south as Aotearoa New Zealand and the edge of the Antarctic Convergence. Swordfish are likely the fastest fish in the world. According to one study they can clock a dizzying eighty miles per hour (130 km/h) when fully underway. Further research suggests that, in addition to the reduction of turbulence due to their body, fin shape, and sword, their skin texture and a special lubricating film of oil excreted out of their foreheads

Broadbill swordfish (*Xiphias gladius*) slashing through prey.

might also help them attain their exceptional speeds. Females grow much larger than males, with individuals topping 1,190 pounds (540 kg) and growing over fourteen feet long (4.5 m). The swords of the adults, which can make up more than half their body length, are flattish with sharp edges—unlike marlin or sailfish rostra, whose bills are more circular. These bills are made of bone, an extension of the fish's upper jaw, but they're not an extended tooth like a mammalian tusk of a walrus or narwhal. Swordfish do not try to impale their prey with their bill, as Piccard wrote. They slash sideways at fish and squids, chopping them into pieces that they can gobble up with their small, toothless mouths.

Humans at sea have been on the wrong end of swordfish spears, afraid of having our boat bottoms pierced, since the earliest days of pushing off the shore. At least as early as the ancient Egyptians, people have depicted swordfish in their artwork. In early Polynesian cultures, warriors used swordfish bills as actual and ceremonial weapons. Early Greco-Roman fishers in the Mediterranean apparently hunted swordfish from boats designed to look like the fish themselves, with a pointed prow and a fish-like hull, thinking they were luring the fish to closely approach what the animals thought was just another large swordfish. Oppian in the second century CE wrote how once harpooned, the swordfish responded in pain and often impaled their bills in the boats' hulls. "The xiphias, or, in other words, the sword-fish," relayed Pliny the Elder in c. 79 CE, "has a sharp-pointed muzzle, with which it is able to pierce the sides of a ship and send it to the bottom."

Fishers and mariners have since shared similar stories of hulls impaled by swordfish or other billfish. In early October 1615, for example, the Dutch mariner-explorer William Cornelison Schouten witnessed an event off Sierra Leone, Africa, while on the way to being the first known European to sail around (and name) Cape Horn. The sailors heard a startling noise by the bow and looked over the rail. The sea, Schouten wrote, looked "all red as if a great store of bloud had bin powred into it, whereat he wondred, knowing not what it ment, but afterward he found, that a great fish or a sea monster having a horne had therewith stricken against the ship, with most great strength." After crossing the Atlantic, they hauled the boat out of the water on a beach in Argentina. Schouten and his crew found that seven feet (2.1 m) below the waterline the fish's bill was still stuck through three planks and then into a rib inside the hull.

In the 1800s, whaleships and merchant ships witnessed similar encounters; some ships even had to be taken ashore for repairs to stop the leak. A few museums in Britain have hunks of ship's frames that were struck by billfish, including a plank from 1847, in which the bill, likely of a marlin in this case, thrust through a four-inch (10 cm) plank, split through an interior timber, and then left its bill projected a further six inches (15.2 cm) inside the hull of the ship. (The vessel was named, ironically, the *Royal Archer.*) This hazard did not stop sailors from trying to catch these fish for food. The men would also often make scrimshaw out of swordfish bills in a similar way as sailors like Thomas Albro did with whale teeth, wiping ink into etched and stippled designs on the wide, flat surface of the swordfish bill. Meanwhile nineteenth-century commercial fishers for cod and halibut, or those trying to harpoon or hook swordfish and tuna out on the North Atlantic or North Pacific reported swordfish bills thrusting straight through their dories and other small boats.

By the 1900s swordfish "attacks" on boats tapered out with the development of engines and steel hulls. Swordfish lunges at watercraft still occurred, though, in part because fishers, commercial and recreational, continue to hunt *Xiphias gladius.* Over two decades into his eclectic fascination with whale shark sightings, Dr. Eugene W. Gudger, by then curator of fishes at the American Museum of Natural History, published a one-hundred-page study that assembled documented collisions between billfish and boats. He titled this study "The Alleged Pugnacity of the Swordfish and the Spearfishes as Shown by their Attacks on Vessels" (1940). Gudger published all known historical sketches and photographs he could find, including an image of a marlin that in 1939 rammed itself all the way through a small row boat off the coast of the California: the fish's body was fully inside and across the boat with its bill out the other side. The shaft of the sportsman's arrow was still stuck in the fish's head.

Two decades after Gudger's survey, in 1962, a Japanese fishing vessel, the thirty-nine-ton *Genyo Maru*, had to be rescued out at sea when a swordfish speared a hole in the engine room and the fifteen-man crew could not keep up with the leak.

The point of all this is that Piccard's interaction with an aggressive swordfish was hardly unique. Which brings us to the question similar to that of Teddy Seymour and his killer whale: what motivates these swordfish, if not retaliating in pain, to purposefully attack a submarine or ship at the surface? *Xiphias* and other billfish likely use their rostrums to defend themselves against sharks or toothed whales. But do they ever go on the offensive against their predators, as previous naturalists and mariners used to think? Maybe they just mistake a ship's hull for a whale? One explanation might be that they get accidentally stuck when slashing through a school of fish congregating under a slow-moving boat. But this seems unlikely, considering the acute awareness fish have about their surroundings and the sheer, deliberate force and speed with which some of these fish impale themselves into a wood hull.

Back aboard the *Benjamin Franklin*, Jacques Piccard mused as to why a swordfish might attack an underwater submersible. In the twenty-first century, swordfish have even entangled themselves with man-made underwater structures, such as drilling platforms or underwater pipes. In the science fiction novel *Twenty Thousand Leagues Under the Sea* (1870), Jules Verne wrote of a ten-foot long swordfish occasionally stabbing at the glass window of his fantastical, yet prescient submarine, the *Nautilus*, a scene that foreshadows Piccard's real-life experience. Perhaps there might be some sort of aggression or defense response to large man-made hulls, which could be mistaken for larger predators. The *Alvin* pilots had wondered if it was the light from the submersible, a reflection of the fish in the window, or maybe even a submersible's similarity, its "eye," to a deep sea squid?

So, yes, swordfish often attack submarines, boats, ships, and pipelines. They always have, and they seemingly always will. Just don't get too stuck, don't blindly stab in the dark, as to the why.

"Decidedly," wrote Jacques Piccard, "the psychology of fishes is barely understood; there is an immense and fascinating field here, but difficult to study."

Yellow-Bellied Sea Snake

In that famous poem of fantasy and horror, "The Rime of the Ancient Mariner," which Mary Brewster notably did *not* mention in her travels with the albatross, a gnarly old sailor wanders from town to town telling his spellbinding story of a supernatural voyage. As punishment for shooting an albatross with an arrow off Cape Horn, magical spirits hold the mariner hostage all the way into the middle of the Pacific Ocean. At one point he is by himself, with no wind or any drinking water, surrounded only by his undead shipmates sprawled around the decks. One moonlit night, just when it seems that he was either going to die of thirst or go insane, he sees some creatures in the water:

> Beyond the shadow of the ship,
> I watched the water-snakes:
> They moved in tracks of shining white,
> And when they reared, the elfish light
> Fell off in hoary flakes.
> Within the shadow of the ship
> I watched their rich attire:
> Blue, glossy green, and velvet black,
> They coiled and swam; and every track
> Was a flash of golden fire.

The ancient mariner is overwhelmed by the beautiful vision of these snakes slithering through the "golden fire," the marine bioluminescence. So he blesses the snakes. This act suddenly frees him from the dark spell. Underwater spirits now hurry his ship home to England.

This isn't a true story, of course, but there are indeed snakes in the ocean. About eighty species of water snakes, better known today as sea snakes, include several species of kraits (which spend a bit of their time on land). Sea snakes swim on the surface of tropical coastal waters in the Pacific and Indian Oceans. They are marine reptiles just like sea turtles, crocodiles, and the marine iguana. Sea snakes breathe air, have scales, and drink fresh water from rain at the surface and other sources by the coast. The true sea snakes are so well adapted to life in the ocean that they do not even go ashore to give birth to their live young. They can dive under the surface for hours on one breath. Evolved for a marine existence, sea snakes have no eyelids, and they have flaps

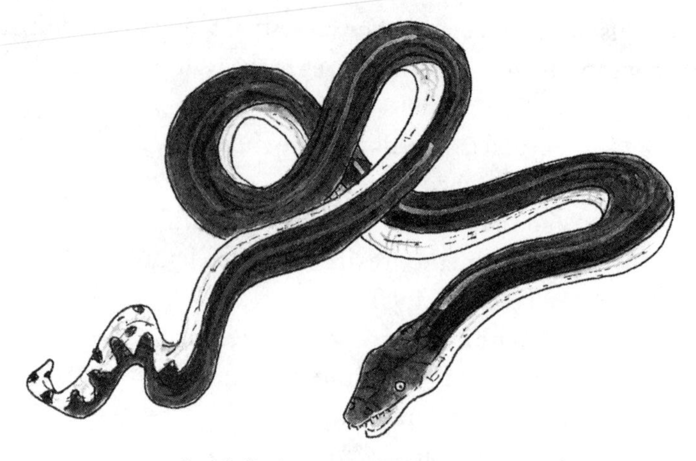

Yellow-bellied sea snake (*Hydrophis platurus*).

over their nostrils to keep water out of their one long lung that stretches nearly the entire length of their body, supplying oxygen as well as helping to regulate their buoyancy under the surface. Sea snake tails are flattened at the end, acting like a rudder when they swim. In order to keep their bodies clear of algae and fouling organisms, such as barnacles, and to slough off dead skin, sea snakes tie themselves in a knot and wipe themselves clean, similar to the famous strategy of hagfish.

Closely related to the poisonous taipan and cobras, sea snakes can inject a neurotoxin with their fangs that attacks the muscles of its prey. Sea snake venom is in fact among the most potent and deadly of any of the snakes, used to kill primarily small fish, crabs, and eels. In humans, a sea snake bite, depending on the species and age of the snake, can cause paralysis, respiratory failure, and often death.

Nearly all sea snakes prefer to live near coastlines, but one species, the yellow-bellied sea snake (*Hydrophis platurus*), also swims around in the open ocean. They display combinations of longitudinal stripes of black, bluish-black, yellow, and brown, but there are some individuals that are entirely yellow. This species is by far the most widely distributed of the sea snakes, found off Africa's Cape of Good Hope, throughout the Indian Ocean and Indonesia, and all the way across the Pacific, from Aotearoa New Zealand up to southern Siberia and across to the coasts of Baja California and Peru. Mariners have observed slicks of thousands of yellow-bellied sea snakes on the surface at once. The snakes slither among patches of foam and floating debris, as they capture fish that like to congregate under shady sticks and drifting seaweeds. The migrations and reproductive behavior of the yellow-bellied sea snakes, however, remain barely known. The largest of these sea snakes are the females; they can grow as long as a person's arm. Unlike most other snakes, yellow-bellied females give birth to from only two to six live young at a time.

For his poem, Coleridge was probably imagining yellow-bellied sea snakes. He had not sailed on the Pacific Ocean himself, but he had read plenty of sailor's journals and stories by the time he completed the "Rime of the Ancient Mariner" in 1798. Records show Coleridge had taken notes on sea snakes, including one account by the pirate-naturalist William Dampier, who wrote of his 1699 voyage to the coast of Australia, explaining, "In the Sea we saw . . . Abundance of Water-Snakes of several Sorts and Sizes."

Coleridge also learned about sea snakes from an account of a voyage by Sir Richard Hawkins that talks about a sick sea, giving diseases to the sailors: "In the yeere of our Lord 1590 . . . many moneths becalmed, the Sea thereby being replenished with seuerall sorts of gellies and formes of Serpents, Adders and Snakes, Greene, Yellow, Blacke, White, and some partie-coloured, whereof many had life, being a yard and a halfe, or two yards long. And [the sailors] could hardly draw a Bucket of Water, cleare of some corruption withall." (As it turns out, there are no sea snakes of any species anywhere in the Atlantic, nor even any land snakes in the Atlantic archipelago of the Azores where he recorded this; so Hawkins's observation here is a bit of a mystery in hindsight. Maybe eels?)

Published in the same year as the "Ancient Mariner," Royal Navy officer James Colnett's account of a British voyage to the South Pacific mentions sea snakes. Colnett explained that off the coast of Peru, "sea snakes were also in great plenty, and many of the crew made a pleasant and nutritious meal of them." A few weeks later, after surviving a near fatal attack by an "alligator" (likely a crocodile) when at anchor on Coiba Island off Panama, Colnett wrote of his belief that the sea snakes in the region were harming the local fish populations. They found in the sea snakes' stomachs undigested fish "of a size that credulity itself would almost refuse to believe."

In the mid-1800s American whalers observed sea snakes in the Pacific, too, and our whaler-artist Thomas L. Albro drew on his sperm whale tooth a long "Sea Serpant" with a flicking forked tongue and stripes along the bottom, looking a bit like some varieties of yellow-bellied sea snakes. Did Albro and his shipmates eat sea snakes? Or bless them? Maybe both.

Today marine biologists know that sea snakes are the most diverse group of marine reptiles in the global ocean and are crucial to saltwater ecosystems. Yet sea snakes are susceptible to fishing gear, coastal pollution, and a loss of coral reef habitats. Predators on sea snakes include sharks, eagles, and seabirds, as well as some humans, such as within communities in Thailand, Vietnam, Japan, Taiwan, and Australia, where fishers capture sea snakes to send to food markets or to sell for their skin. Vietnamese fishers in the Gulf of Thailand, for example, will reach barehanded and even step barefooted into bins filled with hundreds of captured, live sea snakes. The men do occasionally get bitten, for which they prefer to avoid modern hospitals but use instead a piece of horn from the endangered rhino, believing that it helps heal the wound. Records are hard to compile, but it's reasonable to believe that at least a few people die each year as they harvest these sea snakes along the coasts of India and Southeast Asia.

One of the morals of the "Rime of the Ancient Mariner," relevant both in 1798 and even more so in our twenty-first century, is that being kind to animals benefits humans as well. In the poem, bad things happen to the sailor when he kills the majestic albatross. But then he is only truly saved when he blesses the glittering sea snakes, an unlikely animal to elicit gushing sympathy. Few people today would be likely to display a "Save the Sea Snakes" sticker or cuddle a stuffed sea snake toy at home. The ancient mariner, however, teaches us to love "all things both great and small." This kindness, among other results, might someday save you from thirsty zombie shipmates.

Zooplankton

We've arrived at the final story, ready to tie back up at the dock, completing the voyage of this ocean bestiary.

Hopefully Thomas L. Albro would be proud, or at least intrigued about how our knowledge and relationship with the ocean has changed so dramatically since his days at sea and his time helping his local school. Through the eyes of people representing a range of time periods, professions, classes, nationalities, races, ethnicities, genders, and ages, I've told stories that included each of the animals he drew in his scrimshaw bestiary, the ones he engraved onto a sperm whale tooth over several days, perhaps in between doing the officer's laundry or boiling rice while he sat on the deck of his whaleship in the South Pacific in the 1830s. Albro etched his animal drawings with both whimsy and an eye for accuracy.

Over a century later, an American nature writer in 1941 crafted what I believe to be the most artful and significant of ocean natural histories in the English language. Hers was an ocean bestiary too, in its way, so it is fitting that we nod to her masterpiece to conclude this humble one, especially around a story featuring zooplankton.

She grew up on a farm in western Pennsylvania, writing about, exploring, and observing the natural world from the earliest age. She earned scholarships and financial aid and her parents took out loans and sold some of their land so their daughter could attend the Pennsylvania College for Women, what is now Chatham University. Here she majored in English, then switched to biology.

After graduating, she earned a six-week scholarship to study marine biology in Woods Hole, Massachusetts. She made her way to New York City and boarded a steamboat that traveled overnight through Long Island Sound, her first ever glimpse of salt water at the age of twenty-two. She had her brown hair pulled back in a bun.

In the early morning, the boat steamed past the mouth of Narragansett Bay and then came to the dock in New Bedford, Massachusetts. Here Rachel Carson boarded a smaller ferry over to Woods Hole, where for the very first time she was able to explore seaside beaches, dip into the intertidal zone, and examine nets filled with fish and invertebrates dragged up from the ocean bottom. She immersed herself in a library devoted to marine study and learned from working scientists and fishers about what was actually happening beneath the surface. In other words, her summer in Woods Hole was an epiphany for her life's work.

She had already enrolled in a master's program that fall at

Johns Hopkins University, where for her thesis she studied catfish anatomy. Carson was unable to continue on for her PhD, however, because this was during the Great Depression. She needed to help support her family, and there was little funding for marine scientists, particularly if they were women. So she got a job as a writer and editor for the US Fisheries Bureau. Sometimes at night she wrote articles on the fisheries and marine life of the Chesapeake Bay for the *Baltimore Sun*, signing her articles "R. L. Carson," in order to be taken as seriously as a male author might be.

One day in 1935 she came into her supervisor's office to discuss an introduction to a brochure she had written. When her boss returned the draft, he explained that it was too lyrical for a government publication. Perhaps she should submit it to *The Atlantic* magazine?

After some time, she did submit the essay. It was accepted with the title "Undersea," in which she wrote carefully of the driver of all marine life in the ocean, the plankton: "The ocean is a place of paradoxes. It is the home of the great white shark, two-thousand-pound killer of the seas, and of the hundred-foot blue whale, the largest animal that ever lived. It is also the home of living things so small that your two hands might scoop up as many of them as there are stars in the Milky Way. And it is because of the flowering of astronomical numbers of these diminutive plants, known as diatoms, that the surface waters of the ocean are in reality boundless pastures."

In "Undersea" Carson teaches about these photosynthesizing microscopic plants, and then the tiny organisms that feed on them, drifting toward the surface of all seas, known collectively as plankton, "the wanderers."

The diatoms and tiny marine plants are known more specifically as phytoplankton. Their photosynthetic activity, what biol-

Zooplankton that spend their lives as drifters: clockwise around a copepod (*Calanus* sp.) with the two long antennae and a big-eyed hyperiid amphipod (*Hyperia* sp.), are a hydromedusa sea jelly (*Eucheilota* sp.), a lucifer shrimp (*Belzebub* sp.), an arrow worm (*Sagitta* sp.), and a pteropod, or "sea butterfly" (*Limacina* sp.). Each of these is less than 0.5 inch long (~1.3 cm).

ogists call primary productivity, provides the basis of all marine life in the ocean. All the tiny animals that feed on these plants are now known collectively as zooplankton (secondary productivity). Zooplankton include the copepods, the most abundant animals on the planet, both in terms of sheer numbers and their combined weight. Zooplankton also include the early stages of marine fish and of invertebrates before they settle on a permanent substrate or before they grow large enough to physically control their own migrations.

In "Undersea" Carson writes, "Many of the fishes, as well as the bottom-dwelling mollusks and worms and starfish, begin life as temporary members of this roving company, for the ocean cradles their young in its surface waters. The sea is not a solicitous foster mother. The delicate eggs and fragile larvæ are buffeted by storms raging across the open ocean and preyed upon by diminutive monsters, the hungry glassworms and comb jellies of the plankton."

Abalones, *Architeuthis dux*, dolphinfish, electric rays, flying fish, halibut, Juan Fernández crawfish, Louisiana shrimp, octopuses, paper nautiluses, pilot fish, pyrosomes, quahogs, silver kings, teredo shipworms, tuna, urchins, the sea jellies *Velella* and man-of-war, and even *Xiphias gladius* all begin their life cycles as tiny zooplankton.

Rachel Carson was not the first to recognize or describe in awe the significance of phytoplankton and zooplankton for all life on our watery planet. Human hunters and voyagers for millennia had observed whale sharks, baleen whales, and seabirds—like Mother Carey's chickens, noddies, and penguins—feeding on surface plankton. Indigenous mariners and fishers opened up the stomachs of these birds, along with those of fish, seals, and whales, in which they found enormous gulps of small crustaceans, eggs, larvae, and marine worms. As early as the 1600s, people were peering through microscopes at zooplankton in samples of seawater from along the coast and then later from the open ocean aboard ships.

By the early 1800s, mariners such as the English whaler-naturalist William Scoresby Jr., connected their understanding of plankton to their Christian faith, extolling the "profusion of life in a region so remote from the habitations of men!" Scoresby

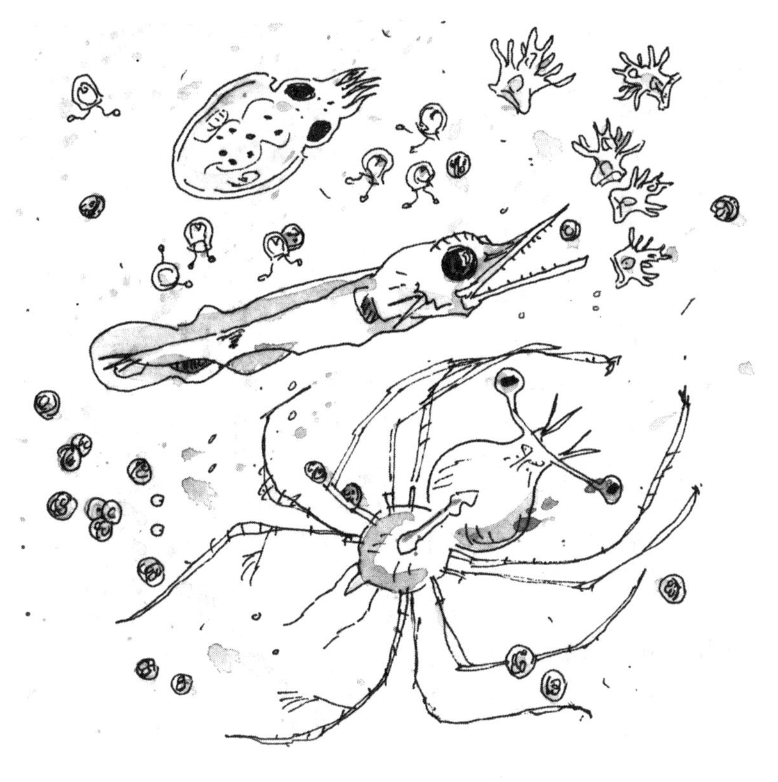

A few animals in early stages as zooplankton: *Velella* medusas around a common octopus paralarva, *Octopus vulgaris* (top left); larval urchins, *Lytechinus variegatus* (top right), larval swordfish, *Xiphias gladius*, ~0.3 inches, 8 mm (middle), phyllosoma of Juan Fernández crawfish, *Jasus frontalis* (bottom), and fish eggs scattered.

had sailed multiple voyages in the far North Atlantic and into the Arctic Circle hunting right whales and bowheads. He, like the Inuit millennia before him, understood how the largest of marine mammals feed on the very smallest invertebrates, "producing a dependent chain of animal life, one particular link of which being destroyed, the whole must necessarily perish." In his *An Account of the Arctic Regions* (1820), Scoresby included a full-page illustration of zooplankton—a bestiary collage of copepods, amphipods, comb jellies, and the floating snails known as pteropods. People on other whaleships, such as Thomas Albro and young Minnie Lawrence, observed these huge patches of krill, copepods, and other zooplankton—what whalers called "brit"—to help them find right and other baleen whales. And sailing the South Pacific in the 1830s at the same time as Albro, the young Charles Darwin, when not harshing on noddies, used buckets to scoop up seawater and towed behind his ship a makeshift plankton net in order to capture and identify plankton. He too wrote of large swaths of krill, what the sealers, he said, simply called "whale food."

A century after Darwin's voyage on the *HMS Beagle*, Rachel Carson's essay "Undersea" appeared in *The Atlantic*. The essay was well-received. Soon she completed her first book, an extended version of the essay, which she titled *Under the Sea-Wind*, published in 1941. This was before underwater film documentaries and before the word *ecology*, the study of relationships between species and habitats, was in wide use. In Carson's 1940s, there was overfishing—there had been for centuries—but the widespread use of plastics in the homes of people around the world had only just begun, and plastics were not yet used to make the fishers' nets or lines. There were as yet no enormous factory trawlers, and, though she was watching this start to change, the majority of the Earth's ocean remained beyond the reach of draggers for fish and beyond the reach of drills for oil.

In *Under the Sea-Wind*, her ocean bestiary in lyrical prose, Carson included no human dialogue, no marine biologists, and very few humans at all—only a few fishers—and only once does she place the reader inside the head of one of these humans, a young person who is out there fishing early in his career, wondering what his prey is actually doing down below the surface: the very question she tries to answer throughout her story. Carson tries to aggregate in *Under the Sea-Wind* all human knowledge at the time about marine and coastal life in and around the western North Atlantic Ocean. She follows two sanderlings, one mackerel and one eel, as these animals migrate through nearly every marine ecosystem of the northwestern Atlantic and through the stages of life—from egg to adult to reproduction to death. She shows a Darwinian world in the sea, one that is unsentimental about birth and death and one in which the element of chance is as crucial as fitness to an individual's ability to live, reproduce, and pass on its genes. With very little anthropomorphism, Carson did her best to evoke in her ocean bestiary—given the limits of human language and knowledge of animal sentience in the 1940s—what might be the experiences and perceptions of seabirds, fish, and marine invertebrates. Carson's writing about zooplankton specifically was the most daring of her writing experiments in *Under the Sea-Wind*. Carson describes the lives of seething sea jellies, of translucent pteropods, of transparent crab larvae molting and transforming toward adulthood, and of leaf-shaped eel larvae, transparent, invisible in the water save only two dark pricks of the developing eyes. Carson writes of the reality that nearly all of the trillions of zooplankton individuals are "doomed to die" quickly, eaten by some slightly larger organism, such as a comb jelly or a tiny fish or a baby green turtle. In one chapter, Carson even takes the reader day by day through the development of an individual mackerel (*Scomber scombrus*). We watch a single fertilized mackerel egg floating out at sea, a fetal fish with a yolk sac. Its eyes and backbone grow rapidly, the larval fish ever and always helpless and in danger. After a week, the mackerel egg hatches, less than a twentieth of an inch (1.3 mm) long, smaller than a pinhead, a minuscule fishlet whose gills and heart and vision continue to develop hour by hour as it "floated

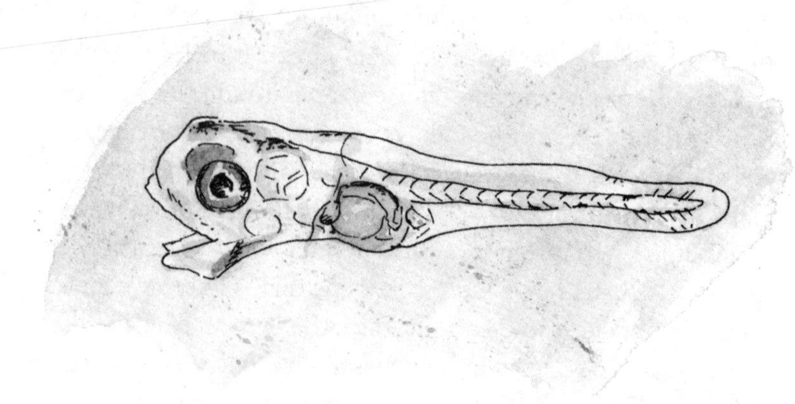

A transparent Atlantic mackerel (*Scomber scombrus*) larva with yolk sac, about 190 hours after hatching and about ⅛" long (~4.2 mm).

where the sea carried him, now a rightful member of the drifting community of the plankton."

Rachel Carson went on to write two more books about the ocean: *The Sea Around Us* (1951) and *The Edge of the Sea* (1955). In both of these she included descriptions and discussions of the importance of zooplankton, but not as imaginatively or in such close focus as in *Under the Sea-Wind*. All three of her ocean books were bestsellers that forever influenced Western global thinking about the ocean, both for marine scientists and everyone else interested in the sea. Then, as Carson was slowly dying from cancer, she completed *Silent Spring* (1962), which exposed and explained to the public the dangers of pesticide use to all life. She is most often identified today with this final work, one of the critical sparks that ignited the modern environmental movement, yet Rachel Carson's love was always for the ocean, and her favorite book was always her first, her beautiful bestiary *Under the Sea-Wind*.

Carson had observed the effects of sea-level rise and global warming and had seen hints of the impacts of industrial pollution and nuclear waste in the ocean, but the scientific research about carbon emissions was only just emerging at the time of her death, just short of her fifty-seventh birthday. Carson was spared learning of the true impacts and extent of deforestation, the burning of fossil fuels, major oil spills, the impacts and persistence of microplastics floating within the plankton layer, and—also just around the chronological corner—the relentless industrial overharvesting of the seas at nearly all trophic levels. Still, though, as Marilyn Nelson would do nearly eighty years later with her poem "Octopus Empire," Carson found hope in scale, in our ocean planet's endurance, in a panned-out vision of geologic time, ending *Under the Sea-Wind* with the waves washing away mountains and human life: "the places of its cities and towns would belong to the sea."

Rachel Carson believed and teaches us still that the inhabitants of the global ocean, from abalones to orcas and zooplankton, whose intelligence and emotional lives remain mostly unknown to us, and the wild waters of the sea in which they have lived for millions of years before human arrival, are on the whole, although unprepared for our onslaught, still astoundingly resilient. Perhaps there is indeed some hope in the perspective that if, for our blink of human time, we are able right now to drastically reduce our individual and collective consumption and to lessen our ecological impact, that just maybe the diverse, vast, carbon-absorbing ocean and its brave, beautiful, and brilliant beasts will be the force that brings some healing.

We might then, at our next meeting at sea, be worthy to look that seabird in the eye.

Acknowledgments

Many of these stories and illustrations were published in different versions in the magazine *Sea History* as part of a quarterly column titled "Animals in Sea History," for which I'm indebted to editor and friend Dee O'Regan, who regularly and generously gave help, expertise, and support. A few other stories found their way in different forms in my previous books, and the story about Matthew Henson's relationship with walruses was told as part of an article for *Oceanus* magazine in August, 2021, coauthored with David Anderson, with editorial help from Evan Lubofsky, Kiara Royer, and Ned Schaumberg.

The illustrations were done with a Rapidograph pen and watercolor paint on hot pressed paper. When I relied directly on a particular photograph or another's illustration, I've cited this source in the bibliography. Erin Greb created the world map.

Thank you to Marilyn Nelson for permission to include here an excerpt from her poem "Octopus Empire," first published online in 2019 with the Academy of American Poets.

Each story benefited from so many people who offered their time and expertise, sometimes for multiple stories. Thank you to experts and friends, including David Anderson, Mary Benouameur, Jim Carlton, the crew of *Catch It!*, Ella Cedarholm, Mea Cook, Sharika Crawford, Laurie Deredita, Bryan Donaldson, Michael Dyer, Ron Eby, Julian Finn, Stuart Frank, Anne Birgitte Gotfredsen, Jefferson Hinke, Wendy Kitchell, Todd McLeish, Neville Peat, Paula Peters, Hugh Powell, Justin Richard, Krystal Rose, Jeff

Schell, Wonseob Song, Elisa Goya Sueyoshi, Fern Van Sant, Kerstin Wasson, Sabina Wilhelm, Jan Witting, Shreya Yadav, and Alex Zapata.

Thank you to decades of students and colleagues at Williams-Mystic and the Sea Education Association, and, as ever, to Alison O'Grady at Williams College, and to Paul O'Pecko and Maribeth Belinski for help with images, books, and articles. Maya Anderson provided invaluable research information on Thomas Albro. Dominick Leskiw gave great editorial feedback on the prose and art in a latter draft, as did Laura Bost and Seth King, who both also helped to improve the illustrations.

Thank you to the entire team at the University of Chicago Press, especially editor Karen Darling, who brought on the project and helped shape and edit the book, as well as editorial assistant Fabiola Enríquez, production editor Beth Ina, designer Kevin Quach, wise and careful copyeditor Nick Murray, and, once again, energetic and thoughtful publicist Nick Lilly. Helen Rozwadowski and two peer reviewers helped refine and shape the book. It's an honor to be an early part of this Oceans in Depth series.

This being a bestiary, I should give, too, a pat on the back to the black lab of luxury, Lola the dog, truly an author-illustrator's best friend. Most importantly, thank you to my spouse Lisa and to our child Alice, artist of the fish on p. 186 and to whom this book is dedicated. They both have been so supportive and served as readers and the most helpful of consultants.

Selected Bibliography

INTRODUCTION

Frank, Stuart M. *Ingenious Contrivances, Curiously Carved: Scrimshaw in the New Bedford Whaling Museum*. Boston: David R. Godine, 2012.

Albro, Thomas. "Engraved Sperm Whale Tooth." Mystic Seaport Museum, no. 1981.40, Mystic, CT.

ABALONE

Carratura, Vincenzo Acampora. "An Interview with a Haenyeo Female Diver." *Korea.net* (February 15, 2017): https://www.korea.net/NewsFocus/People/view?articleId=144131.

Sang-Hun, Choe. "Hardy Divers in Korea Strait, 'Sea Women' Are Dwindling." *New York Times* (March 29, 2014): https://www.nytimes.com/2014/03/30/world/asia/hardy-divers-in-korea-strait-sea-women-are-dwindling.html.

See, Lisa. *The Island of Sea Women*. New York: Scribner, 2019.

Song, Wonseob. "Sustainability of the Jeju *Haenyeo* Fisheries Systems in the Context of Globally Important Agricultural Heritage System (GIAHS)." *Sustainability* 12, no. 9 (2020): 1–18.

Vileisis, Anne. *Abalone: The Remarkable History and Uncertain Future of California's Iconic Shellfish*. Corvallis: Oregon State University Press, 2020.

UNESCO. "Culture of Jeju Haenyeo (Women Divers)." United Nations Educational, Scientific and Cultural Organization: Intangible Cultural Heritage, Session 11 (2016): https://ich.unesco.org/en/RL/culture-of-jeju-haenyeo-women-divers-01068.

Young-Sun, Heo, Yoo Chul-in, Joo Kang-hyun, and Lee Jin-joo. "Special Feature: Iconic Female Divers of Jeju." [multiple articles], *Koreana* 28, no.

2 (Summer 2014): 4–25, https://issuu.com/the_korea_foundation/docs/koreana_summer_2014__english__215792b9ea6138.

ARCHITEUTHIS DUX (GIANT SQUID)

Ellis, Richard. *The Search for the Giant Squid*. New York, Penguin, 1999.

Harvey, Moses. "How I Discovered the Great Devil-Fish." *Wide World Magazine* 2, no. 12 (March 1899): 732–40.

Kubodera, Tsunemi, and Kyoichi Mori, "First-ever Observations of a Live Giant Squid in the Wild." *Proceedings of the Royal Society B* 272 (2005): 2583–86.

Paxton, C. G. M. "Unleashing the Kraken: On the Maximum Length in Giant Squid (*Architeuthis* sp.)." *Journal of Zoology* 300, no. 2 (2016): 82–88.

Verrill, A. E. *The Cephalopods of the Northeastern Coast of America*. Part 1. Washington: Government Printing Office, 1882.

BELUGA WHALE

Bushnell, Kelly B. "Looking at Leviathan: The First Live Cetaceans in Britain." In *Ecocriticism and the Anthropocene in Nineteenth-Century Art and Visual Culture*, edited by Maura Coughlin and Emily Gephart, 178–91. New York: Routledge, 2020.

Dollman, J. C. "The Whale at the Westminster Aquarium." *Graphic* (8 June 1878): 559–60.

Kovacs, Kit M., et al. "Introduction: Beluga Whales (*Delphinapterus leucas*): Knowledge from the Wild, Human Care and TEK." *Polar Research* special cluster, vol. 40 (2021): 1–4.

Lee, Henry. *The White Whale*. London: RK Burt, 1878.

O'Corry-Crowe, Gregory M. "Beluga Whale." In *Encyclopedia of Marine Mammals*, 3rd ed., edited by Bernd Würsig, J. G. M. Thewissen, and Kit M. Kovacs, 93–96. London: Academic Press, 2018.

CHINSTRAP PENGUIN

Cox, Lynne. *Swimming to Antarctica: Tales of a Long-Distance Swimmer*. New York: Alfred A. Knopf, 2004.

Hinke, Jefferson, et al. "Individual Variation in Migratory Movements of Chinstrap Penguins Leads to Widespread Occupancy of Ice-Free Winter Habitats over the Continental Shelf and Deep Ocean Basins of the Southern Ocean." *PLoS One*, 14, no. 12: https://doi.org/10.1371/journal.pone.0226207.

Klieber, Shannon Henry. "Extreme Swimmer Lynne Cox on Her Icy, Record-Breaking Open Water Swim." *Wisconsin Public Radio* (December 20, 2019): https://www.wpr.org/extreme-swimmer-lynne-cox-her-icy-record-breaking-open-water-swim.

Martínez, I., D. A. Christie, F. Jutglar, and E. F. J. Garcia. "Chinstrap Penguin (*Pygoscelis antarcticus*), 1.0." In *Birds of the World*, edited by J. del Hoyo, et al. Ithaca, NY: Cornell Lab of Ornithology, 2020: birdsoftheworld.org/bow/species/chipen2/cur/introduction.

Strycker, Noah, et al. "A Global Population Assessment of the Chinstrap Penguin (*Pygoscelis antarcticus*)." *Scientific Reports* 10, no. 19474 (2020): doi.org/10.1038/s41598-020-76479-3.

DOLPHINFISH

Byron, Lord. *The Complete Works of Lord Byron*. Edited by John Galt. Paris: Baudry's European Library, 1835.

Dana, Richard Henry, Jr. *Two Years before The Mast*. New York: Penguin, 1986.

Falconer, William. *The Poetical Works of William Falconer*. Edited by Thomas Park. London: Stanhope Press, 1809.

Heyerdahl, Thor. *Kon-Tiki: Across the Pacific by Raft*. Translated by F. H. Lyon. New York: Pocket Books, 1973.

Kroll, Gary. *America's Ocean Wilderness: A Cultural History of Twentieth-Century Exploration*. Lawrence: University Press of Kansas, 2008.

Pepperell, Julian. *Fishes of the Open Ocean: A Natural History & Illustrated Guide*. Chicago: University of Chicago Press, 2010.

ELECTRIC RAY

Clark, Eugenie. *Lady with a Spear*. New York: Harper & Brothers, 1953.

Finger, Stanley, and Marco Piccolino. *The Shocking History of Electric Fishes: From Ancient Epochs to the Birth of Modern Neurophysiology*. Oxford, Oxford University Press, 2011.

Kroll, Gary. *America's Ocean Wilderness: A Cultural History of Twentieth-Century Exploration*. Lawrence: University Press of Kansas, 2008.

Rutger, Hayley. "Remembering Mote's 'Shark Lady': The Life and Legacy of

Dr. Eugenie Clark." *Mote Marine Laboratory and Aquarium* (March 5, 2015): mote.org/news.

FLYING FISH

Darwin, Charles. *On the Origin of Species*. New York: Penguin, 2009.

Joffe, Gabrielle. "Flying Fish." *Sea History*, no. 132 (Autumn 2010): 36–37.

McCunn, Ruthanne Lum. *Sole Survivor*. San Francisco: Design Enterprises, 1985.

Pepperell, Julian. *Fishes of the Open Ocean: A Natural History & Illustrated Guide*. Chicago: University of Chicago Press, 2010.

Reiger, George. "Ben Franklin's Dolphins and Flyingfish," *Underwater Naturalist: The Bulletin of the American Littoral Society* 25, no. 1 (2000): 10–12.

FRIGATEBIRD

Audubon, John James. *Ornithological Biography*. Vol. 3. Edinburgh: Adam and Charles Black, 1835.

Brinckley, Edward S., and Alec Humann. "Frigatebirds." In *The Sibley Guide to Bird Life and Behavior*, edited by Chris Elphick, John B. Dunning Jr., and David Allen Sibley, 167–69. New York: Alfred A. Knopf, 2009.

Columbus, Christopher. *The Log of Christopher Columbus*. Translated by Robert H. Fuson Camden, ME: International Marine, 1992, 69.

"COP15 Kiribati Side Event—Song of the Frigate Te Itei," posted by Marc Honore, United Nations Framework Convention on Climate Change, COP 15 (Copenhagen: 9 December 2009): https://www.youtube.com/watch?v=xOcMLWVNIms&t=51s.

Diamond, A. W., and E. A. Schreiber. "Magnificent Frigatebird (*Fregata magnificens*), 1.0." In *Birds of the World*, edited by A. F. Poole and F. B. Gill. Ithaca, NY: Cornell Lab of Ornithology, 2020: https://birdsoftheworld.org/bow/species/magfri/cur/introduction.

Melville, Herman. *Typee: A Peep at Polynesian Life*. Edited by Harrison Hayford, Hershel Parker, and G. Thomas Tanselle. Evanston and Chicago: Northwestern University Press and The Newberry Library, 1968.

Stimson, J. Frank. *Songs and Tales of the Sea Kings: Interpretations of the Oral Literature of Polynesia*. Salem, MA: Peabody Museum, 1957.

Stone, Gregory S. and David Obura. *Underwater Eden: Saving the Last Coral Wilderness on Earth*. Chicago: University of Chicago Press, 2013.

Whitman, Walt. "To the Man-of-War-Bird." In *The Sea Is a Continual Miracle: Sea Poems and Other Writings by Walt Whitman*, edited by Jeffrey Yang, 198. Hanover, NH: University Press of New England, 2017.

GRAMPUS

"Four Girls on a Schooner." *New York Times* (August 23, 1896): 2.

Goode, George Browne, et al. "The Grampuses or Cowfishes." In *The Fisheries and Fisheries Industries of the United States*, Section 1, 13–14. Washington: Government Printing Office, 1884.

Hutching, Gerard. "Dolphins—Humans and Dolphins: The Story of Pelorus Jack." In *Te Ara, The Encyclopedia of New Zealand* (September 1, 2015): https://teara.govt.nz/en/photograph/4696/the-story-of-pelorus-jack.

Kipling, Rudyard. *Captains Courageous*. New York: Signet Classic, 1981.

NOAA Fisheries. "Risso's Dolphin." NOAA Species Directory: www.fisheries.noaa.gov/species/rissos-dolphin.

Reeves, Randall R., Brent S. Stewart, Phillip J. Clapham, James A. Powell, and Pieter A. Folkens. *Guide to Marine Mammals of the World*. New York: Knopf, 2002.

GREEN TURTLE

Carr, Archie F.. "The Passing of the Fleet." *AIBS Bulletin* 4, no. 5 (October 1954): 17–19.

Crawford, Sharika D. *The Last Turtlemen of the Caribbean: Waterscapes of Labor, Conservation, and Boundary Making*. Chapel Hill: University of North Carolina Press, 2020.

Exquemelin, Alexander O. *The Buccaneers of America*. Translated by Alexis Brown. Mineola, NY: Dover, 2000.

Jackson, J. B. C. "Reefs Since Columbus." *Coral Reefs* 16 (1997): S23–S32.

Nowack, Gerald. "Egypt, Red Sea, Green Sea Turtle (*Chelonia mydas*) and Pilot Fishes (*Naucrates ductor*)" [photograph]. *Westend61*: https://www.westend61.de/en/imageView/GNF01123/egypt-red-sea-green-sea-turtle-chelonia-mydas-and-pilot-fishes-naucrates-ductor

Perrine, Doug. *Sea Turtles of the World*. Stillwater, MN: Voyageur Press, 2003.

Spotila, James R. *Sea Turtles: A Complete Guide to Their Biology, Behavior, and Conservation*. Baltimore: Johns Hopkins University Press, 2004.

van der Zee, Jurjan P. "Population Recovery Changes Population Composition at a Major Southern Caribbean Juvenile Development Habitat for the Green Turtle, *Chelonia mydas*." *Scientific Reports* 9, no. 14392 (October 7, 2019): https://www.nature.com/articles/s41598-019-50753-5.

GUANAY CORMORANT

Duffy, David C. "The Guano Islands of Peru: The Once and Future Management of a Renewable Resource." *BirdLife Conservation Series* 1 (1994): 68–76.

Durfee, Nell, and Ernesto Benavides. "Holy Crap! A Trip to the World's Largest Guano-Producing Islands." *Audubon.org* (April 27, 2018): www.audubon.org/news/holy-crap-trip-worlds-largest-guano-producing-islands.

Murphy, Robert Cushman. *Bird Islands of Peru: The Record of a Sojourn on the West Coast*. New York: G. P. Putnam's Sons, 1925.

Nelson, J. Bryan. *Pelicans, Cormorants and Their Relatives*. Oxford: Oxford University Press, 2005.

HALIBUT

Bolster, W. Jeffrey. *The Mortal Sea: Fishing the Atlantic in the Age of Sail*. Cambridge, MA: Belknap Press of Harvard University Press, 2012.

Grasso, Glenn M. "What Appeared Limitless Plenty: The Rise and Fall of the Nineteenth-Century Atlantic Halibut Fishery." *Environmental History* 13, no. 1 (January 2008): 66–91.

"Halibut." *Marine Stewardship Council*. https://www.msc.org/what-you-can-do/eat-sustainable-seafood/fish-to-eat/halibut#.

Homer, Winslow. *The Fog Warning*. Museum of Fine Arts Boston: https://artsandculture.google.com/asset/the-fog-warning/GAEcgEvHorgctQ?hl=en.

Morton, Thomas. *The New English Canaan*. Edited by Charles Francis Adams Jr. Boston: Prince Society, 1883.

Provost, Paul Raymond. "Winslow Homer's *The Fog Warning*: The Fisherman as Heroic Character." *American Art Journal* 22, no. 1 (Spring 1990): 20–27.

HORSE

Rogers, John G. *Origins of Sea Terms*. Mystic, CT: Mystic Seaport, 1985.

"The Dead Horse Festival." *Harper's Weekly* 26, no. 1351 (Nov. 11, 1882): 717–18.

Williams, James H. "A Son of Ishmael." *The Independent* 61 (Nov. 29, 1906): 1267–71.

———. *Blow the Man Down! A Yankee Seaman's Adventures Under Sail*. Edited by Warren F. Keuhl. New York: E. P. Dutton, 1959.

ISURUS OXYRINCHUS (MAKO SHARK)

Beegel, Susan F. "A Guide to the Marine Life in Ernest Hemingway's *The Old Man and the Sea*." *Resources for American Literary Study* 30 (2005): 236–315.

———. "The Monster of Cojímar: A Meditation on Hemingway, Sharks, and War." *Hemingway Review* 34, no. 2 (Spring 2015): 9–35.

Ebert, David A., Sarah Fowler, and Marc Dando. *A Pocket Guide to Sharks of the World*. Princeton, NJ: Princeton University Press, 2015.

Hemingway, Ernest. *The Old Man and the Sea*. New York: Scribner, 1986.

———. *El Viejo y El Mar*. Translated by Lino Novas Calvo. New York: Vintage Español, 2003.

Schmaltz, C. S. Rafinesque. *Caratteri Di Alcuni Nuovi Generi e Nuove Specie Di Animali e Plante Della Sicilia*. Palermo: Sanfilippo, 1810.

JUAN FERNÁNDEZ CRAWFISH

Eddy, Tyler D., Jonathan P. A. Gardner, and Alejandro Pérez-Matus. "Applying Fishers' Ecological Knowledge to Construct Past and Future Lobster Stocks in the Juan Fernández Archipelago, Chile." *PLoS One* 5, no. 11 (2010): doi.org/10.1371/journal.pone.0013670.

Heyerdahl, Thor. *Aku-Aku: The Secret of Easter Island*. Chicago: Rand McNally, 1958.

Howell, John. *The Life and Adventures of Alexander Selkirk*. Edinburgh: Oliver and Boyd, 1829.

Muñoz, Alex. "The Ocean-Saving Power of Local Communities." *MacArthur Foundation: Perspectives* (February 4, 2021): https://www.macfound.org/press/perspectives/the-ocean-saving-power-of-local-communities.

Rogers, Woodes. *A Cruising Voyage Round the World . . . Begun in 1708, and finish'd in 1711*. London: Cassell, 1928.

Selcraig, Bruce. "The Real Robinson Crusoe." *Smithsonian* (July 2005): https://www.smithsonianmag.com/history/the-real-robinson-crusoe-74877644/.

Wahle, R., A. MacDiarmid, M. Butler, and A. Cockcroft. "Juan Fernandez Rock Lobster, *Jasus frontalis*." The IUCN Red List of Threatened Species (2011): https://www.iucnredlist.org/ja/species/169948/6690436.

Walpole, Frederick. *Four Years in the Pacific, in Her Majesty's Ship "Collingwood."* vol. 1. London: Richard Bentley, 1850.

KILLER WHALE

Esteban, Ruth, et al. "Killer Whales of the Strait of Gibraltar, an Endangered Subpopulation Showing a Disruptive Behavior." *Marine Mammal Science* (May 18, 2022): 1–11.

Ford, John K. B. "Killer Whale." In *Encyclopedia of Marine Mammals*. 3rd ed. Edited by Bernd Würsig, J. G. M. Thewissen, and Kit M. Kovacs, 531–37. London: Academic Press, 2018.

Ross, Herman. "Teddy Seymour." Atlantic Creole: Black Folk Don't Sail, Blog Post 33 (May 13, 2014): https://herossea.blogspot.com/2014/05/atlantic-creole-black-folk-dont-sail_14.html?q=Seymour.

Seymour, Teddy. "'Love Song' Heads for the Red Sea." *Blue Moment: UK Sailing Cruising Directory* (n.d.): https://www.bluemoment.com/seymour.html.

Whitehead, Hal, and Luke Rendell. *The Cultural Lives of Whales and Dolphins*. Chicago: University of Chicago Press, 2015.

LOUISIANA SHRIMP

Field, Martha R. *Louisiana Voyages: The Travel Writings of Catharine Cole*. Edited by Joan and Jack McLaughlin. Jackson: University Press of Mississippi, 2006.

Kaplan Levenson, Laine. "Dancing the Shrimp Dry: How Chinese Immigrants Drove Louisiana Seafood." *Gravy Podcast* (September 8, 2016): https://www.southernfoodways.org/gravy/dancing-the-shrimp-dry-how-chinese-immigrants-drove-louisiana-seafood-gravy-ep-45/.

"White Shrimp." NOAA Fisheries: https://www.fisheries.noaa.gov/species/white-shrimp.

MOTHER CAREY'S CHICKENS (STORM-PETRELS)

Audubon, John James. *Audubon at Sea*, edited by Christoph Irmscher and Richard J. King. Chicago: University Press of Chicago, 2022.

Chichester, Francis. *Gipsy Moth Circles the World*. New York: Coward-McCann, 1967.

Lawrence, Mary Chipman. *The Captain's Best Mate: The Journal of Mary Chipman Lawrence on the Whaler* Addison, *1856–1860*. Edited by Stanton Garner. Hanover, NH: Brown University Press/University Press of New England, 1966.

Leland, Charles Godfrey. *Songs of the Sea and Lays of the Land*. London: Adam and Charles Black, 1895.

Melville, Herman. "The Encantadas." *The Piazza Tales and Other Prose Pieces*. Edited by Harrison Hayford, Alma A. MacDougall, G. Thomas Tanselle, et al. Evanston and Chicago: Northwestern University Press and the Newberry Library, 1987.

Olney, Derek, and Paul Scofield, *Albatrosses, Petrels, & Shearwaters of the World*. Princeton, NJ: Princeton University Press, 2007.

NEW ZEALAND SEA LION

Chilvers, B. Louise. "New Zealand Sea Lion." In *Encyclopedia of Marine Mammals*, 3rd ed., edited by Bernd Würsig, J. G. M. Thewissen, and Kit M. Kovacs, 635–37. London: Academic Press, 2018.

Department of Conservation *Te Papa Atawhai*. "New Zealand sea lion/rāpoka/whakahao": https://www.doc.govt.nz/nature/native-animals/marine-mammals/seals/new-zealand-sea-lion.

Frans, Veronica F., et al., "Integrated SDM Database: Enhancing the Relevance and Utility of Species Distribution Models in Conservation Management." *Methods in Ecology and Evolution* 13, no. 1 (January 2022): 243–61.

New Zealand Sea Lion Trust, www.sealiontrust.org.nz.

Peat, Neville. *Coasting: The Sea Lion and the Lark*. Dunedin: Longacre Press, 2001.

Peck, George Washington. *Melbourne, and the Chincha Islands; with sketches of Lima, and A Voyage Round the World*. New York: Charles Scribner, 1854.

Raynal, F. E. *Wrecked on a Reef; or, Twenty Months Among the Auckland Isles, A True Story*. London: T. Nelson and Sons, 1874.

NODDY

Audubon, John James. *Audubon at Sea*, edited by Christoph Irmscher and Richard J. King. Chicago: University Press of Chicago, 2022.

Darwin, Charles. *Voyage of the* Beagle. Edited by Janet Browne and Michael Neve. London: Penguin, 1989.

Harrison, Peter. *Seabirds: An Identification Guide*. Boston: Houghton Mifflin, 1985.

Lewis, David. *We, the Navigators: The Ancient Art of Landfinding in the Pacific*. Honolulu: University of Hawai'i Press, 1975.

Stephens, J. F. "Aves." In *General Zoology or Systematic Natural History*, edited by Charles Shaw. Vol. 8, pt. 1. London: Thomas Davison, 1825.

OCTOPUS

Bryner, Jeanna. "Octlantis: See Photos of Tight-Knit Gloomy Octopus Communities." *Live Science* (October 30, 2017): https://www.livescience.com/60803-octlantis-photos-gloomy-octopus.html.

Hugo, Victor. *The Works of Victor Hugo: Toilers of the Sea*. Translated by Isabel F. Hapgood. Vol. 1. New York: Thomas Y. Crowell, 1888.

Klingel, Gilbert C. *The Ocean Island: Inagua*. New York: Dodd, Mead, 1940.

Nelson, Marilyn. "Octopus Empire." Academy of American Poets (2019): https://poets.org/poem/octopus-empire. This excerpt reprinted with her kind permission.

Nolen, R. Scott. "Scary Smart: Of All the Species Aquatic Veterinarians Treat, the Octopus Is Among the Most Unique." *American Veterinary Medical Association* (June 30, 2015): www.avma.org/javma-news/2015-07-15/scary-smart.

Pliny the Elder. *Natural History: A Selection*. New York: Penguin, 2004.

Stokstad, Erik. "Scientists Discover an Underwater City Full of Gloomy Octopuses." *Science* (13 September 2017): doi: 10.1126/science.aap9543.

OTTER

Harvey, Jim. "Those That Used the Hill Before Us," Moss Landing Marine Laboratories, August 7, 2015, https://mlml.sjsu.edu/2015/08/13/those-that-used-the-hill-before-us/.

"Marine Mammals: IV. Sea Otter." Monterey Bay National Marine Sanctuary: https://montereybay.noaa.gov/sitechar/mamm4.html.
Moss, Madonna L. "Did Tlingit Ancestors Eat Sea Otters? Addressing Intellectual Property and Cultural Heritage through Zooarchaeology." *American Antiquity* 85, no. 2 (2020): 202–221.
McLeish, Todd. *Return of the Sea Otter: The Story of the Animal that Evaded Extinction on the Pacific Coast*. Seattle: Sasquatch Books, 2018.
Ohlone Cotanoan Esselen Nation, official tribal website: http://www.ohlonecostanoanesselennation.org/index.html.
Salomon, Anne K., Kii'iljuus Barb J. Wilson, Xanius Elroy White, Nick Tanape Sr., and Tom Mexsis Happynook. "First Nations Perspectives on Sea Otter Conservation in British Columbia and Alaska: Insights into Coupled Human—Ocean Systems." In *Sea Otter Conservation*, edited by Shawn E. Larson, et al., 301–31. London: Elsevier Science & Technology, 2015.

PAPER NAUTILUS

"Argonaut Buoyancy." Museums Victoria (2010): www.youtube.com/watch?v=EgISnrhSAmQ.
De Villepreux-Power, Mme Jeannette. *Observations Physiques sur Le Poulpe de L'Argonauta Argo*. Paris: Imprimerie Charles de Mourgues Frères: 1856.
Finn, Julian K., and Mark D. Norman. "The Argonaut Shell: Gas-mediated Buoyancy Control in a Pelagic Octopus." *Proceedings of the Royal Society B* 277 (19 May 2010): 2967–2971.
Wood, William. *Zoography; or, the Beauties of Nature Displayed*. Illustrated by William Daniell. Vol. 2. London: Cadell and Davies, 1807.

PARROT

Alex Foundation. "Alex." Laboratory and Staff of Irene Pepperberg: https://alexfoundation.org/the-birds/alex/.
Boehrer, Bruce Thomas. *Parrot Culture: Our 2,500-Year-Long Fascination with the World's Most Talkative Bird*. Philadelphia: University of Pennsylvania Press, 2004.
Cordingly, David. *Under the Black Flag*. New York: Harcourt Brace, 1995.
Dampier, William. *Dampier's Voyages*. Edited by John Masefield. Vol. 2. London: E. Grant Richards, 1906.

Hagseth, Megan C. "Seadogs and Their Parrots: The Reality of 'Pretty Polly.'" *Mariner's Mirror* 104, no. 2 (May 2018): 135–52.

Johnson, Charles. *Middle Passage*. New York: Scribner, 2015.

Stevenson, Robert Louis. *Treasure Island*. New York: Bantam, 1981.

PILOT FISH

Bennett, Frederick. *Narrative of a Whaling Voyage Round the Globe*. Vol. 2. London: Richard Bentley, 1840.

Correia, João Pedro Santos, et al. "Capture, Husbandry and Long-term Transport of Pilotfish, *Naucrates ductor* (Linnaeus, 1758), by Sea, Land and Air." *Environmental Biology of Fishes* 101, no. 6 (June, 2018): 1039–52.

Ebert, David A., Sarah Fowler, and Marc Dando. *A Pocket Guide to Sharks of the World*. Princeton, NJ: Princeton University Press, 2015.

Melville, Herman. "The Maldive Shark." In *Published Poems*, edited by Robert C. Ryan, Harrison Hayford, Alma MacDougall Reising, and G. Thomas Tanselle. Evanston and Chicago: Northwestern University Press and The Newberry Library, 2009.

Whitefield, George. *A Journal of a Voyage from London to Savannah in Georgia*. London: W. Strahan, 1743.

QUAHOG

Ahtone (Ahtoneharjo-Growingthunder), Tahnee M., and Elizabeth James-Perry. *Curating Indigeneity*. Podcast, episode 2 (September 6, 2018): https://anchor.fm/curating-indigeneity/episodes/Episode-2-e24tcj.

Bradley, James W. "Re-visiting Wampum and Other Seventeenth-century Shell Games." *Archaeology of Eastern North America* 39 (2011): 25–51.

MacKenzie, Clyde L., et al. "Quahogs in Eastern North America: Part 1, Biology, Ecology, and Historical Uses." *Marine Fisheries Review* 64, no. 2 (July 2002): 1–55.

Peters, Paula. "Why Wampanoag Truths and Traditions Are So Crucial to the Mayflower Story." The Box (Plymouth), Mayflower 400 UK (2021): https://www.youtube.com/watch?v=9z6mnN8zyIU.

———. "The Making of the Wampum Belt." The Box (Plymouth), Mayflower 400 UK (2021): https://www.youtube.com/watch?v=OAxWoSYqOUw.

"Two Row Wampum, Gä·sweñta'." Onondaga Nation: https://www.onondaganation.org/culture/wampum/two-row-wampum-belt-guswenta/.

RIGHT WHALE

Austin, Heather. "Southern Right Whale (*Eubalaena australis*) Five-Year Review: Summary and Evaluation." National Marine Fisheries Service. Silver Spring, MD: Office of Protected Resources (March 2021): 1–68.

Bowles, M. E. "Some Account of the Whale-Fishery of the N. West Coast and Kamschatka." *The Polynesian* 2, no. 20 (October 4, 1845): 82–83.

Carroll, Emma. "Genome and Satellite Technology Reveal Recovery Rates and Impacts of Climate Change on Southern Right Whales." *The Conversation* (October 29, 2020): theconversation.com.

Jackson, J. A., N. J. Patenaude, E. L. Carroll, and C. Scott Baker. "How Few Whales Were There after Whaling? Inference from Contemporary mtDNA Diversity." *Molecular Ecology* 17 (2008): 236–51.

"Journal of a Voyage Aboard the *Merrimac*, 1844–47, and the *General Williams*, July–Nov. 1852," edited by Laurie M. Deredita. Frank L. McGuire Maritime Library of the New London Maritime Society: https://mcguirelibrary1998.Omeka.net.

Laist, David W. *North Atlantic Right Whales: From Hunted Leviathan to Conservation Icon*. Baltimore: Johns Hopkins University Press, 2017.

Lawrence, Mary Chipman. *The Captain's Best Mate: The Journal of Mary Chipman Lawrence on the Whaler* Addison, *1856–1860*, edited by Stanton Garner. Hanover, NH: Brown University Press/University Press of New England, 1966.

SEA COW

Anderson, Paul K., and Daryl P. Domning. "Steller's Sea Cow." In *Encyclopedia of Marine Mammals*, 3rd ed., edited by Bernd Würsig, J. G. M. Thewissen, and Kit M. Kovacs, 935–38. London: Academic Press, 2018. Reconstruction illustration by Uko Gorter.

Littlepage, Dean. *Steller's Island: Adventures of a Pioneer Naturalist in Alaska*. Illustrated by Judy Shimono. Seattle: The Mountaineers Books, 2006.

Reeves, Randall R., Brent S. Stewart, Phillip J. Clapham, James A. Powell, and Pieter A. Folkens. *Guide to Marine Mammals of the World*. New York: Knopf, 2002.

Steller, Georg Wilhelm. "De Bestiis Marinis, or, The Beasts of the Sea (1751)." Translated by Walter Miller and Jeannie Emerson Miller. Edited by Paul Royster. *Faculty Publications, University of Nebraska–Lincoln Libraries*, Paper 17 (1751): 1–89.

———. *Journal of a Voyage with Bering, 1741–1742*. Edited by O. W. Frost. Translated by Margritt A. Engel and O. W. Frost. Stanford, CA: Stanford University Press, 1988.

SEA PICKLE (PYROSOME)

Péron, M. "Mémoir: Sur la nouveau genre Pyrosoma." *Annales du Muséum National d'histoire naturelle* 4 (1804): 437–46. A portion was translated, along with Frederick Bennett's account, in "Tunicata," in *The Cyclopædia of Anatomy and Physiology*, edited by Robert B. Todd, vol. 4, part 2, 1227–30 (London: Longman, et al., 1849–1852).

Sea Education Association. "SSV Robert C Seamans." https://sea.edu/ships-and-crew/ssv-robert-c-seamans/, and https://sea.edu/blog/bogs-from-the-robert-c-seamans/ (blog).

SILVER KING

Adams, A., et al. "Tarpon, *Megalops atlanticus*." The IUCN Red List of Threatened Species (2019, rev. errata 2020): https://www.iucnredlist.org/species/191823/174796143.

Dean, Harry, and Sterling North. *Umbala: The Adventures of a Negro Sea-Captain in Africa and on the Seven Seas in His Attempts to Found an Ethiopian Empire [The Pedro Gorino]*. Winchester, MA: Pluto Press, 1989.

Luo, Jiangang, et al. "Migrations and Movements of Atlantic Tarpon Revealed by Two Decades of Satellite Tagging." *Fish and Fisheries* (2019): 1–29.

Pepperell, Julian. *Fishes of the Open Ocean: A Natural History & Illustrated Guide*. Chicago: University of Chicago Press, 2010.

TEREDO SHIPWORM

Carlton, James T. "Shipworm." In *The Oxford Encyclopedia of Maritime History*, edited by John B. Hattendorf, 3:694–96. Oxford: Oxford University Press, 2007.

Columbus, Christopher. "Letter of Columbus on the Fourth Voyage." In *The Northmen, Columbus, and Cabot, 985–1503*, edited by Julius E. Olson and Edward Gaylord Bourne, 403–7. New York: Charles Scribner's Sons, 1906.

Nelson, Derek Lee. "The Ravages of Teredo: The Rise and Fall of Shipworm in US History, 1860–1940." *Environmental History* 21, no. 1 (January 2016): 100–24.

Yonge, C. M. and T. E. Thompson. *Living Marine Molluscs*. London: Collins, 1976.

TROPICBIRD

Brinkley, Edward S., and Alec Humann, "Tropicbirds." In *The Sibley Guide to Bird Life and Behavior*, edited by Chris Elphick, John B. Dunning Jr., and David Allen Sibley, 151–53. New York: Alfred A. Knopf, 2001.

Davison, Ann. *My Ship Is So Small*. New York: William Sloane, 1956.

Lee, D. S., and M. Walsh-McGee. "White-tailed Tropicbird (*Phaethon lepturus*), 1.0." In *Birds of the World*, edited by S. M. Billerman. Ithaca, NY: Cornell Lab of Ornithology, 2020: https://birdsoftheworld.org/bow/species/whttro/cur/introduction.

"Felicity Ann." The Community Boat Project. https://communityboats.wordpress.com/programs/felicity-ann.

———. Northwest School of Wooden Boatbuilding: https://www.nwswb.edu/felicityann/.

Palliser, Tony. *Red-tailed Tropicbird*. N.d. Photograph. Christmas Island Tourism Association.

TUNA

Battuta, Ibn. *Ibn Battuta in The Maldives and Ceylon*. Translated by Albert Gray. New Delhi: Asian Educational Services, 2004.

Collette, B. B., et al. "Skipjack Tuna, *Katsuwonus pelamis*." The IUCN Red List of Threatened Species (2021): https://www.iucnredlist.org/species/170310/46644566.

Pepperell, Julian. *Fishes of the Open Ocean: A Natural History & Illustrated Guide*. Chicago: University of Chicago Press, 2010.

Pollon, Christopher. "Tuna's Last Stand." *Hakai* (March 2, 2021): https://www.hakaimagazine.com/features/tunas-last-stand/.

Shadwick, Robert Edward. "How Tunas and Lamnid Sharks Swim: An Evolutionary Convergence." *American Scientist* 93, no. 6 (2005): https://doi.org/10.1511/2005.56.524.

Yadav, Shreya, Ameer Abdulla, Ned Bertz, and Alexander Mawyer, "King Tuna: Indian Ocean Trade, Offshore Fishing, and Coral Reef Resilience in the Maldives Archipelago." *ICES Journal of Marine Science* 77, no. 1 (2020): 398–407.

URCHIN

Aristotle. "Historia Animalium." Edited by D'Arcy Wentworth Thompson. In *The Works of Aristotle*, edited by J. A. Smith, W. D. Ross. Vol. 4. Oxford: Clarendon Press, 1910.

Cang, Voltaire. "The Three Great 'Rare Tastes' in Japanese Culinary History: Sea Cucumber Entrails, Sea Urchin Gonads and Mullet Row." In *Offal: Rejected and Reclaimed Food, Proceedings of the Oxford Symposium on Food and Cookery 2016*, edited by Mark McWilliams, 121–33. London: Prospect Books, 2017.

James, Philip, and Sten Siikavuopio. "A Guide to the Sea Urchin Reproductive Cycle and Staging Sea Urchin Gonad Samples." Nofima and the Norwegian Seafood Research Fund (June 2012): 1–20.

Lawrence, John M. "Edible Sea Urchins: Use and Life-History Strategies." In *Edible Sea Urchins: Biology and Ecology*, 2nd ed., edited by John Miller Lawrence, 1–6. Amsterdam: Elsevier, 2007.

Pontoppidan, Erich. *The Natural History of Norway*. Translated from Danish. London: A. Linde, 1755.

VELELLA AND THE MAN-OF-WAR

Barfield, Peter, "A Mass Stranding of *Velella* (Linnaeus, 1758), the By-the-wind Sailor, North-east Sicily, April 2015." *Bulletin of the Porcupine Marine Natural History Society*, no. 5 (Spring 2016): 26–29.

Francis, Lisbeth. "Sailing Downwind: Aerodynamic Performance of the *Velella* Sail." *Journal of Experimental Biology* 158, no. 1 (1991): 117–32.

MacMichael, Morton, III. *A Landlubber's Log of His Voyage around Cape Horn*. Philadelphia: J. B. Lippincott, 1883.

WALRUS

Bolster, W. Jeffrey. *The Mortal Sea: Fishing the Atlantic in the Age of Sail*. Cambridge, MA: Belknap/Harvard University Press, 2012.

Eegeesiak, Okalik, Eva Aariak, Kuupik V. Kleist. "People of the Ice Bridge: The Future of the Pikialasorsuaq." The Pikialasorsuaq Commission (November 2017): i-C-33, www.pikialasorsuaq.org.

Gotfredsen, Anne Birgitte, Martin Appelt, and Kristen Hastrup. "Walrus History Around the North Water: Human-Animal Relations in a Long-term Perspective." *Ambio* 47, supp. 2 (2018): S193–S212.

Henson, Matthew A. *A Negro Explorer at the North Pole*. New York: Frederick A. Stokes, 1912.

King, Richard J., and David Anderson. "Diverse Voices from Our Maritime Past: Exploring the Value of Historical Marine Observations by Mariners of Color." *Oceanus, the Magazine of Woods Hole Oceanographic Institution*. 19 August 2021: www.whoi.edu/oceanus/feature/diverse-voices-from-our-maritime-past.

Lyderson, Christian. "Walrus." In *Encyclopedia of Marine Mammals*, 3rd ed., edited by Bernd Würsig, J. G. M. Thewissen, and Kit M. Kovacs, 1045–48. London: Academic Press, 2018.

Roberts, Callum. *The Unnatural History of the Sea*. Washington, DC: Shearwater/Island Press, 2007.

WANDERING ALBATROSS

Barwell, Graham. *Albatross*. London: Reaktion, 2014.

Brewster, Mary. *"She Was a Sister Sailor": Mary Brewster's Whaling Journals, 1845–1851*. Edited by Joan Druett. Mystic, CT: Mystic Seaport, 1992.

Coleridge, Samuel Taylor. *The Annotated Ancient Mariner*. Edited by Martin Gardner. New York: Prometheus Books, 2003.

Murphy, Robert Cushman. *Logbook for Grace*. Chicago: Time-Life Books, 1982.

Olney, Derek, and Paul Scofield. *Albatrosses, Petrels, and Shearwaters of the World*. Princeton, NJ: Princeton University Press, 2007.

WHALE SHARK

Clark, Eugenie. "Whale Sharks: Gentle Monsters of the Deep." *National Geographic* 182, no. 6 (December 1992): 121–39.

"Forty-five-foot-long shark (whale?) on trailer; two children on its back." Library of Congress Prints and Photographs Division, Washington DC, ca. 1913. https://www.loc.gov/pictures/resource/cph.3b11709/.

Gudger, E. W. "Natural History of the Whale Shark." *Zoologica: Scientific Contributions of the New York Zoological Society* 1, no. 19 (March 1915): 349–89.

Holmberg, Jason, ed. "Wild Book for Whale Sharks," *Wild Book*: www.whaleshark.org.

Pierce, S. J., and B. Norman. "Whale Shark, *Rhincodon typus*." The IUCN Red List of Threatened Species (2016): https://www.iucnredlist.org/species/19488/2365291.

XIPHIAS GLADIUS (SWORDFISH)

Gudger, E. W. "The Alleged Pugnacity of the Swordfish and the Spearfishes as Shown by their Attacks on Vessels." *Memoirs of the Royal Asiatic Society of Bengal* 12, no. 2 (1940): 215–315.

Hsing-Juin Lee, Yow-Jeng Jong, Li-Min Chang, and Wen-Lin Wu. "Propulsion Strategy Analysis of High-Speed Swordfish." *Transactions of the Japan Society for Aeronautical and Space Sciences* 52, no. 175 (2009): 11.

Oppian. "Halieutica." In *Oppian, Colluthus, Tryphiodorus*, translated by Alexander W. Mair, 200–522. Cambridge, MA: Harvard University Press, 1963.

Pepperell, Julian. *Fishes of the Open Ocean: A Natural History & Illustrated Guide*. Chicago: University of Chicago Press, 2010.

Piccard, Jacques. *The Sun beneath the Sea*. New York: Charles Scribner's Sons, 1971.

Pliny the Elder. *The Natural History of Pliny*. Edited and translated by John Bostock and H. T. Riley, 6:8. London: Henry G. Bohn, 1857.

Schouten, William Cornelison. *The Relation of a Wonderfull Voiage* [sic]. London: Nathanaell Newberry, 1619.

Sokol, Joshua. "Sharks Wash Up on Beaches, Stabbed by Swordfish." *New York Times* (October 27, 2020): https://www.nytimes.com/2020/10/27/science/swordfish-stabbing-sharks.html. Accompanying photograph, copied for this illustration, is by Paulo Oliviera/Alamy.

Videler, John J., et al. "Lubricating the Swordfish Head." *Journal of Experimental Biology* 219 (2016): 1953.

YELLOW-BELLIED SEA SNAKE

Coleridge, Samuel Taylor. "The Annotated Ancient Mariner." Edited by Martin Gardner, 75, 106. New York: Prometheus Books, 2003.

Colnett, James. *A Voyage to the South Atlantic and Round Cape Horn into the Pacific Ocean . . .* London: W. Bennett, 1798.

Dampier, William. *Dampier's Voyages*. Edited by John Masefield, 2:435. London: E. Grant Richards, 1906.

Gopalakrishnakone, P., ed. *Sea Snake Toxinology*. Kent Ridge: Singapore University Press, 1994.

Guinea, M., et al. "Yellow-bellied Sea Snake, *Hydrophis platurus*," The IUCN Red List of Threatened Species (2017): https://www.iucnredlist.org/species/176738/115883818.

Lillywhite, Harvey B., et al. "Why Are There No Sea Snakes in the Atlantic?" *BioScience* 68 (November 10, 2017): 15–24.

Lowes, John Livingston. *The Road to Xanadu: A Study in the Ways of the Imagination*. New York: Vintage Books, 1959.

MacLeish, Kenneth. "Diving with Sea Snakes." *National Geographic* 141, no. 4 (April 1972): 565–78.

Mattison, Chris. *The New Encyclopedia of Snakes*. Princeton, NJ: Princeton University Press, 2007.

Rasmussen, Arne Redstead, et al. "Marine Reptiles," *PLoS One* 6, no. 11 (November 2011): 1–12.

Takacs, Zoltan. "Is Eating Venomous Sea Snakes a Bad Thing?" *National Geographic* (January 9, 2015): https://www.youtube.com/watch?v=Foc4dn9On3E.

ZOOPLANKTON

Berrien, Peter L. "A Description of Atlantic Mackerel, *Scomber scombrus*, Eggs and Early Larvae." *Fishery Bulletin* 73, no. 1 (1975): 186–92.

Carson, Rachel. "Undersea." In *Lost Woods: The Discovered Writing of Rachel Carson*, edited by Linda Lear, 3–11. Boston: Beacon Press, 1998.

———. *Under the Sea-Wind*. New York: Penguin, 2007.

Pollock, Leland W. *A Practical Guide to the Marine Animals of Northeastern North America*. New Brunswick, NJ: Rutgers University Press, 1998.

Scoresby, William, Jr. *An Account of the Arctic Regions*. Vol. 1. Edinburgh: Archibald Constable, 1820.

Index

Page numbers in italics refer to images.